简美 SIMPLE BEAUTY II

居住空间 DWELING SPACE

open 欧朋文化 黄滢 主编

華中科技大學出版社
http://www.hustp.com
中国 · 武汉

现代风尚
告别冷、硬、淡、静
这一季要让你心花怒放

黄滢

当现代主义建筑大师柯布西埃通过作品系统呈现现代建筑核心内容的六条基本原则:

1. 柱支撑结构，而不是传统的墙承力结构;
2. 房子下部留空，形成建筑的六个面，而不是传统的五个面;
3. 房顶不但是平顶结构，而且设计为屋顶平台，作为天台花园，供居住者休闲用;
4. 室内完全敞开的设计，尽量减少用墙面分隔房间的传统方式;
5. 完全没有装饰的立面;
6. 窗户用条形，与建筑本身的承力结构无关，因而窗结构独立。

当密斯的“少即是多”被设计界奉为圭臬,“一句话改变了世界三分之一的城市天际线”(Tom Wolfte 在《从包豪斯到我们的住宅》中说到)，影响力跨越建筑界，一直延伸到产品设计、服装设计等领域，人类已经无法阻挡现代主义成为当代设计的主流。

现代主义的理性、冷静、功能化，一方面简化便利着我们的生活，另一方面却在束缚着个性的张扬。一栋现代主义的建筑，无论是来自太平洋岛上或者都市公园、城市近郊或者森林山谷，看上去都差不多。极简、极酷、极冷静，也可能极不近人情。就像范斯沃斯极不喜欢密斯为她设计的那栋著名的森林住宅，她最后把密斯告上了法庭。她觉得新住宅剥夺了她的个人隐私，好象赤裸裸地暴露在自然中。由此可见现代主义与心理舒适感之间存在一定的矛盾。

好在设计从来都是为人服务的，没什么规则不能打破，没什么束缚不能解脱，没什么领域不能创新。相对建筑的严谨理性，室内设计可腾挪变化的空间要大得多。我们既可以享受现代主义设计的功能强大、条理分明、清爽洁净，又可以把个人的喜好、个性、梦想收纳其中。

这一季我们推出《简美 II》，所有推荐作品在空间结构上保持现代主义风格的敞朗开阔，而在内里的设计与装饰中，更注重的是个性风格的充分表达。每一间住宅都注入主人的品味、审美、爱好、收藏与对生活的热爱。每一间都显得如此风格卓立，与众不同。

《简美 II》一共分为缤纷、形塑、简素三大章节。第一章缤纷，以饱和的色彩搭配、鲜亮的图形组合、明媚的风格特点，带给人开朗、热情、活力的正向能量。谁说现代主义就是

MODERNIST STYLE,
SAY GOODBYE TO COLD, HARDNESS, WATERINESS AND QUIET,
THIS SEASON WOULD BURST YOUR HEART WITH JOY.

一张黑白脸，只要加上色彩的肆意组合、纵情挥洒，就能够创造出空间的万千表情。

第二章形塑，从空间造型的大胆创意，以及定制家具的突破组合，以最具创新的精神，把空间个性表达得淋漓尽致。搬空家具，空间就像一个艺术馆，搬进入住，生活得就像艺术家。

第三章简素，是现代主义空间基本特质的完美呈现。无论是工艺，还是比例，用品和设施都搭配得极精极简极严格。但这并不防碍设计师在个性化上表现出聪明才智。比如 Naka House 黑色外墙，白色内墙，纤细的框架线条，以及开阔的内里空间，都是现代设计的代表之作。而宽厚的玫红沙发，却为空间注入了浪漫的气息。现代主义风格因为极简，对每一处装饰，每一个家具的设计，每一种装饰材质的质感都要求严苛，只有极高品质的产品才能散发出强大的气场，让整个空间熠熠生辉。所以简约不是简陋的代名词，它是品质与品味的试金石，能在极简的空间里做出个性风范，才有能力让繁复的空间井然有序。

《简美 II》让设计回归人性的心理需求，以缤纷多彩的设计，个性化的创意组合，让你感受到激情与梦想火花四溢的强烈撞击。现代主义的新风尚，是告别冷、硬、淡、静，这一季要让你心花怒放。

目录 DIRECTORY///

POPS OF COLORS 缤纷

SHAPING 形塑

SIMPLE AND PURE 简素

POPS OF COLORS

缤纷

无限色彩，情趣万千，绘出童梦生活

UNLIMITED COLORS AND FUN, DRAW CHILDREN' DREAM

Project Name: Brentwood Residence
Design Company: Jamie Bush & Co

项目名称：布伦特伍德居住
设计公司：杰米布什室内设计

色彩与艺术的盛妆舞步
COLORS AND ARTS

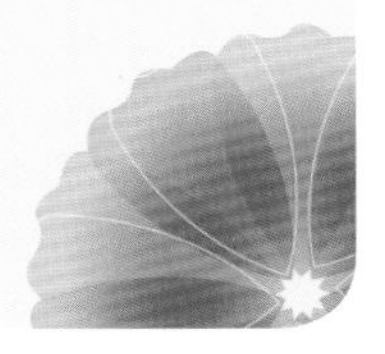

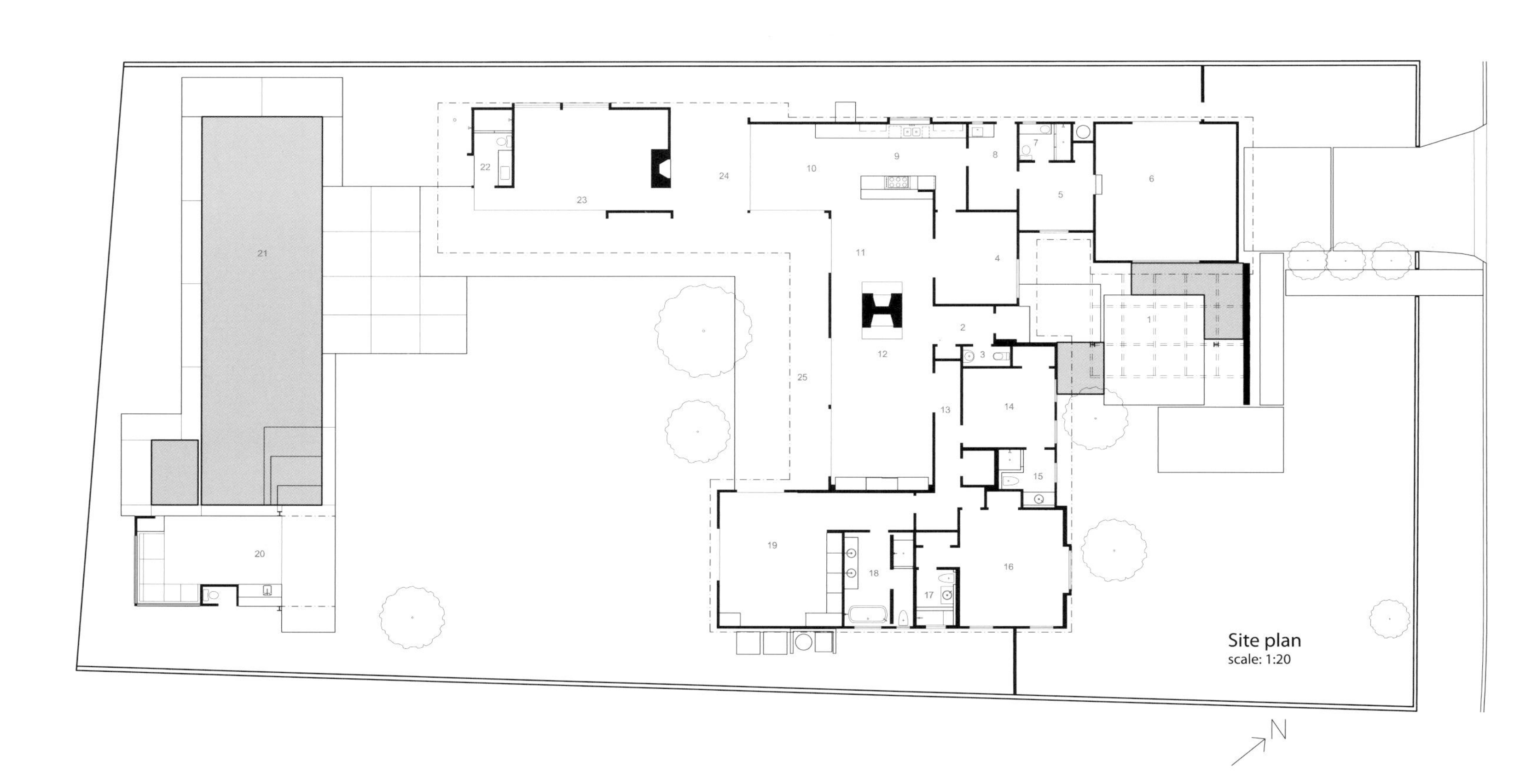
1
2
3
4
5
6
7
8
9
10
11
12
13
14
15
16
17
18
19
20
21
22
23
24
25
Site plan
scale: 1:20
N

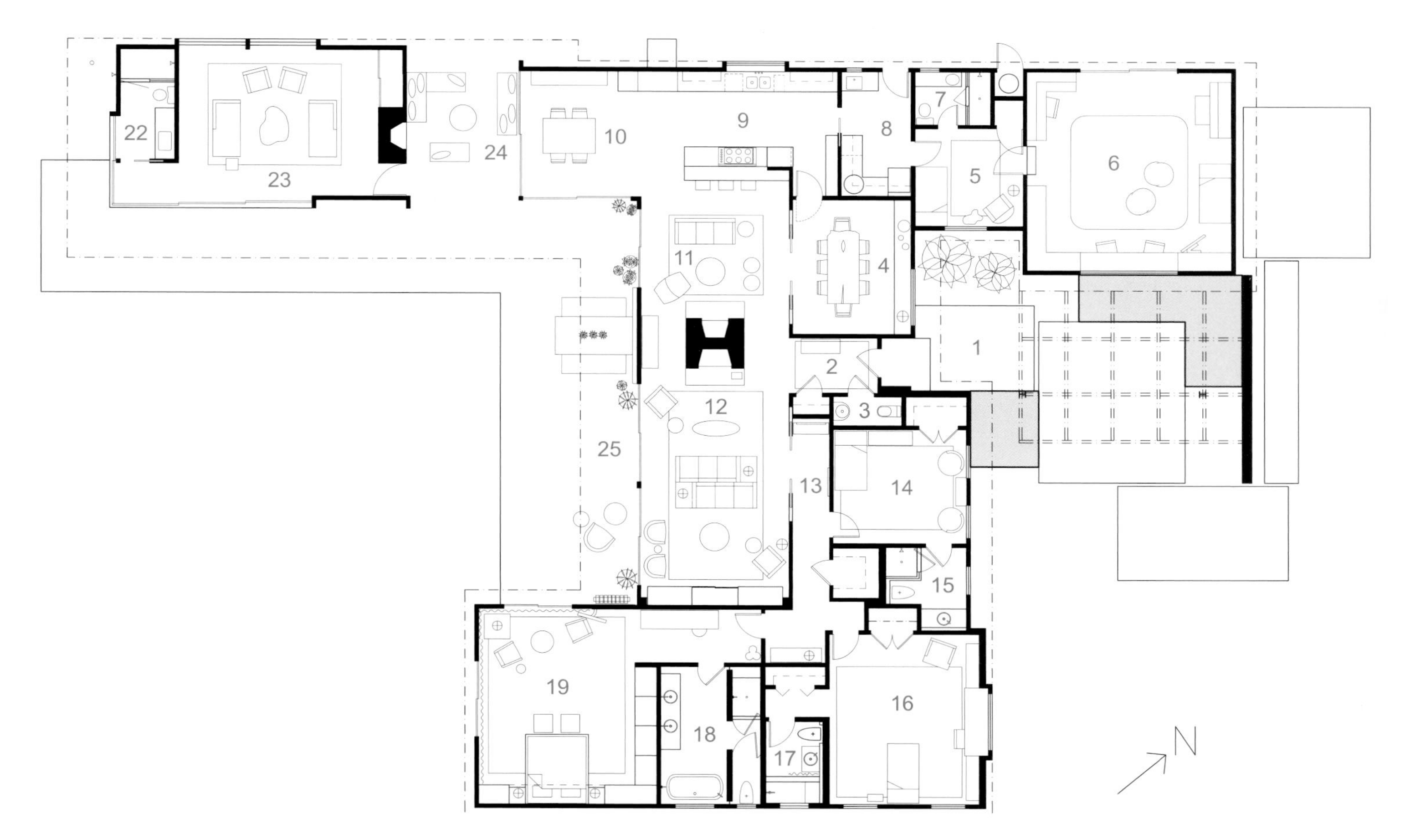

1
2
3
4
5
6
7
8
9
10
11
12
13
14
15
16
17
18
19
22
23
24
25
N

ROSS BLECKNER
MAPPLETHORPE

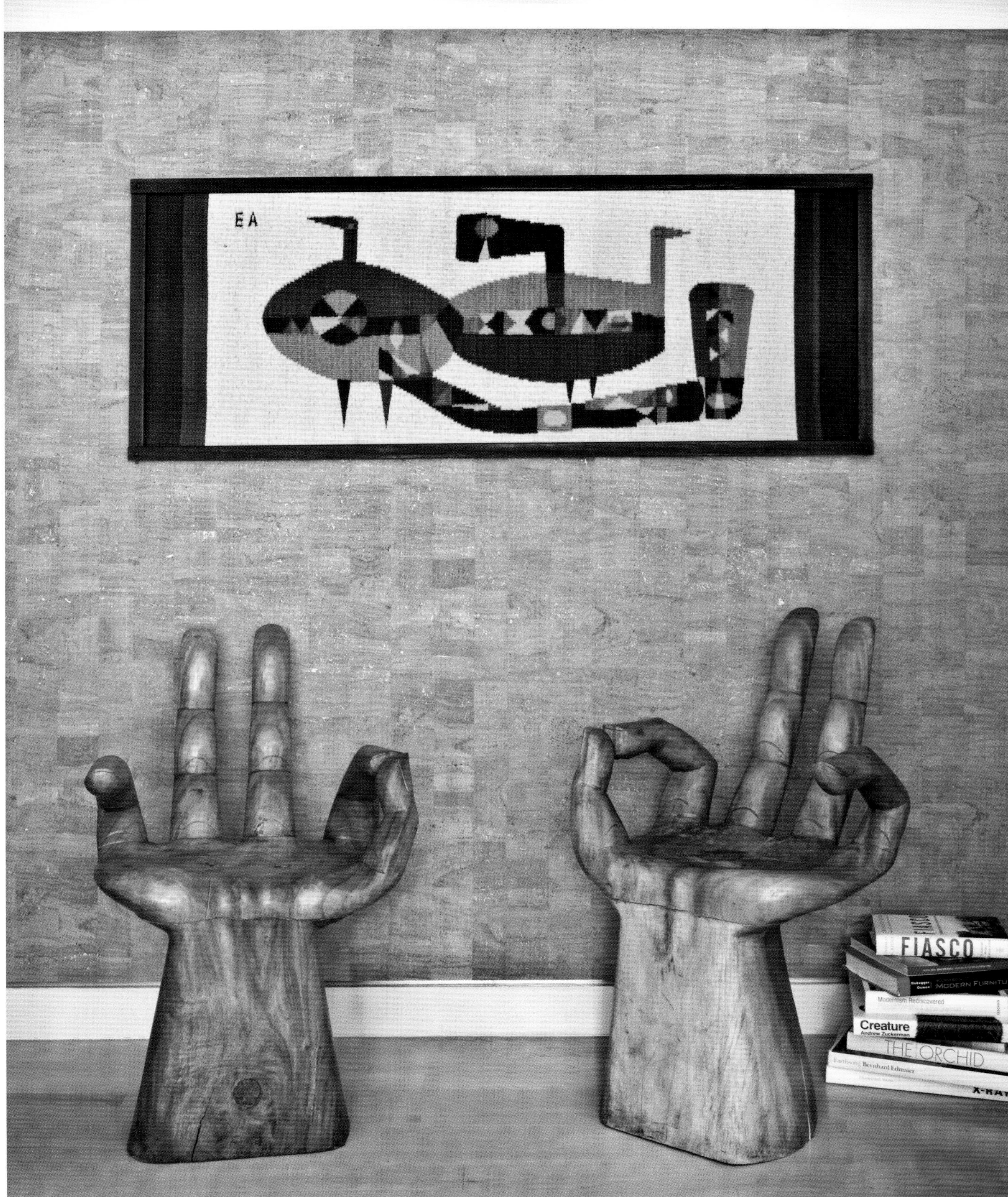
EA
FIASCO
MODERN FURNITURE
Modernism Rediscovered
Creature
Andrew Zuckerman
THE ORCHID
Bernhard Edmaier
X-RAY

本案位于洛杉矶郊外，空间自有一种低调的奢华。其设计由杰米布什承担。

本案放映室、早餐区由布鲁斯设计。外立面、结构设计、泳池及池边建筑的设计则由杰米担当。二位设计师反复探讨，不同设计灵感、理念终汇成立面、结构、内饰、铺陈之和谐。形成了月亮众星捧月之光彩。

业主钟情于中世纪物件。岁月冉冉，时至今日，各物件的色彩或流行，或自然，或保持着自己的超然。三位成年子女各有所好，并各需有自己的生活空间。对于空间设计而言，纷杂的空间功能使命真是个不小的难题。于是早餐区摆上红色、橙色的雷蒙德 Lowey 书柜。而其他空间也自有适合其主题的铺陈，更加朴实的室内摆设，自然的调色板自有局部的流光溢彩点亮。

This mid century modern project by Jamie Bush & Co. is in Brentwood, California, an affluent yet modest suburb of Los Angeles.

Working with architect Bruce Bolander, who first completed the screening room and breakfast area, Jamie addressed the exterior of the house and structural issues inside the house, as well as the pool and pool house. Bruce and Jamie bounced ideas back and forth with each other but also had their own distinct rolls to play – exterior and structural versus interiors, finishes and furnishings.

The owners of the home have a taste for mid century pieces, some with pops of colors and others more natural and neutral. They have three young children and wanted something playful but also adult enough for them to grow into as they mature. They already had the red and orange Raymond Lowey credenza in the breakfast area and a few natural pieces throughout the house. Jamie used that foundation as a starting point and the concept progressed from there. In the process, the focus became demonstrating variations of this theme in each room, using a bright pop of color, which was then grounded by a more natural palette of materials.

Project Name: Casa Micheli
Design Company: Arch. Simone Micheli
Photos: Juergen Eheim
Size: 200 m²

项目名称：米凯利别苑
设计公司：Arch. Simone Micheli
摄影：尤尔根
面积：200 m²

童话糖果屋
HANSEL AND GRETEL

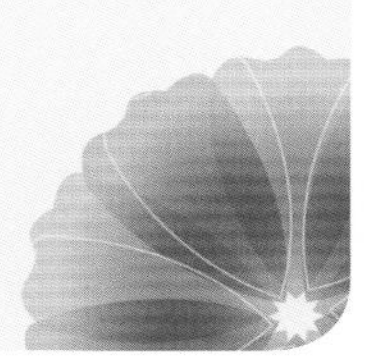

米凯利别苑以环保建材铸就，是豪华美学的超真实写照。19 世纪的量体空间，200 年的华丽穿越，尽显现代、鲜明、生动的都市生活。

全高空间，大型砖质拱门自然界定。别致天花图案，千娇百媚，书写别样风情，引领绿苑宽景。

各空间功能井然有序。稍低的空间，楔形设计，融合了包括厨房在内的服务空间，尤其是楼梯，还宛转引领儿童娱乐空间。服务区眼波流转，天花以国际知名艺术家托马斯•萨拉切诺（Tomas Saraceno）的蜘蛛网状装置为设计灵感，营造了一种优雅、轻盈之气氛。

客厅引发纵向动线行进，一通道与其交叉，成其两翼，一翼为厨房。另一翼为门户界定，壁龛设计与原有量体开窗成垂直态势，空间布局轻松实现新增、原有建筑量体的无缝对接。内置雕塑，通体雪白。

沿动线向前，一全高屏风自天花而下，阻隔着厨房的视线。依屏风组合玻璃台面，或午饭，或工作，彰显生活舒适便利。

游走于精简、激进与时髦之间，几分媚彩，几分梦幻，几分完美，几分纯洁，又有几分童真。

书柜以酸性绿处理，是那么耀眼夺目。镜材家具、卧榻，轻柔如天边的云彩。即便

10 x

原本厚重的背墙，在感染之下，也为空间平添了一种生动，一份活泼。

空间设计主色调一如纯洁的大海，彩色的气泡漂浮在上空。那是孩童特有的纯真、轻快与欢乐。

地体由大尺寸瓷漆瓷砖铺就，亮洁的墙体，超现实色彩的生态皮革装饰，厨房所用漆面，镜面家具合力打造一个浑然雪白的世界。可调试点射灯置于天花假顶，小束灯光一闪一亮，共同簇拥空间之浪漫。

原有建筑量体内纳卧室及一浴。内里空间，纯白色系，极尽建筑设计之精华。家具铺陈如图腾展示，别有风情暗暗升腾。

双人卧室亦梦亦幻。床头一盏宽面圆镜，蓝色背光，状如飞碟。床垫之下，毯状支持，其势如飞。其边角折叠，直至床头板材，并与皮质装饰相互呼应。白色漆面壁橱，装以镀铬把手，绵延至整个墙体。另有大型镜屏依墙而立，呼应着电视的存在；瓷质花瓶恰立于低矮的木质椭圆架上，野趣横生。

主卧，其势和谐，设计考究；另有卧室，乃子女所用，其势欢快。依两墙走是一蛇形解构床，集功能性与美观性为一体。或放玩具，或放书籍，或放寝具。床上自然升起气泡，稚真童趣轻松化解了对面墙面壁橱带来的僵硬质感。漆面书桌、书架，其形圆润，洒下一室温暖。黑色地毯、背光镜面，其形依然圆润。如此圆润的气度内，自有洞天，为孩子创造一个宜嬉戏、学习、生活的万能空间。

两卫浴设计简约且清爽。内里装饰铺陈气质温柔，饰面无瑕。各部件设计，如人体器官，令人惊叹不已。整体环境如灵魂之依托，并以灵魂为导向。都市生活，魅力无穷。繁杂的都市丛林，人来人往，步履匆匆。而进入空间、万事为空，这自是本案设计师的初衷。

The house of Cesar, Roberta and Simone Micheli is a residence made up of 90% eco compatible materials and it is an authentic hyperrealist portrait of the "Ethical Luxury" which is one of the main focuses of his daily architectural searching. It is a dynamic, extremely fresh and vivid intervention taking place in an ancient 1 800 setting which has converted these spaces laden with memories into a new environment capable of hosting meaningful fragments connected with a fast and unstoppable metropolitan life.

In this way the full-height space which breaks the residential distribution rules is divided by a big brick arch, it features an unconventional distribution pattern on the ceiling and large openings revealing the opposite garden.

It hosts the episodes of daily life in succession while a wedge, a low height diagonal volumetric element hosts services and the kitchen. Above this splinter which can be reached by walking the staircase we find a space for children to play marking the end of the whole visual description pattern. A spiderweb, a thread texture, somewhere between an Indian dreamcatcher and a microcosm appeared to Tomas Saraceno in a vision watches over the floor and creates an informal and valuable frame which embellishes the body of the services in an elegant and light way. The longitudinal axis which is triggered by the volumetric development of the living room meets a counterpoint, a minor transept in correspondence with the kitchen, a snow white sculpture marked exclusively by the partition of the doors which forms a niche in the added body resulting in dilating space in the dimension perpendicular to the windows.

All along this transveral dilatation glass tables for lunch and work are located together with a screen coming down the ceiling which can visually separate the kitchen from the rest of the environment.

This informal crystalline, naive, immaculate, antibourgeois, suspended between minimalism and radical chic is torn by ringing and vivid patches of colors.

The bookcase, an acid green highlight, the mirror furniture and couches, which look as soft as pink clouds, as well as the back wall lend liveliness to the overall architecture.

The perception of this space is highly unconventional, visionary where color bubbles floating in a spotless sea generate emotions connected with childish innocence, lightness and joy.

The big size enamelled grès porcelain tiles floor are absolute white, they are as bright as the walls, the surreal ecoleather spherical poufs and the kitchen lacquering, the mirror furniture side surfaces and the bases of the stuffed furniture. The lighting featuring built in adjustable spotlights located in the false ceiling and the lamps with very narrow optics underline and enhance the lyric volumetric and cromatic episodes of this composition.

The bedrooms and another bathroom are located in a transversal partition of the original volume. The bedrooms as well as the rest of the residence are characterized by extreme whiteness and by essential architectural gestures. The furniture is composed by few single totemic elements.

The double bedroom is suspended between dream and reality. A big blue backlit round mirror flies over the bed like a flying saucer. The bed mattress is supported by a flying carpet which forms the leather upholstered structured headboard with its fold. The white lacquered closet covering the whole wall is embellished with essential chromium plated handles and it sets the perfect background for other dreamlike and reassuring environments. The big mirror resting against the wall sports a TV, while the ceramic vases which look as smooth as river stones are put on a low wooden elliptical shelf.

On the one hand the bedroom of Roberta e Simone is harmonious and delicate, on the other hand the room of Cesar is joyful and sparkling. A big yellow snake which represents a deformation of the standard bed unit stretches along the two sides of the room and it becomes the support for toys, books and for the mattress. Soda bubbles come up from the bed and color joyfully the wall opposite the space totally filled by the wall closet. Yellow lacquered round shaped desk and shelves together with the white ecoleather cylindrical pouff, the circular black carpets and the round backlit mirror complete this lively and childfriendly space which is ideal for playing, studying and living in a highly stimulating environment.

The two bathrooms follow a simple and formally clean pattern and feature few essential elements having soft shapes and spotless finishing. Many pieces of furniture are similar to moulds of the human body, to hollow shapes capable of holding it; the whole environment represents the mould of the soul, the case without whom the soul would be like a snail without its shell" . Therefore this house originated from the mould of a lively, dynamic, metropolitan future oriented soul. It is a contemporary, intriguing, unpredictable mould which has the unmistakable signature of Architect Simone Micheli forged.

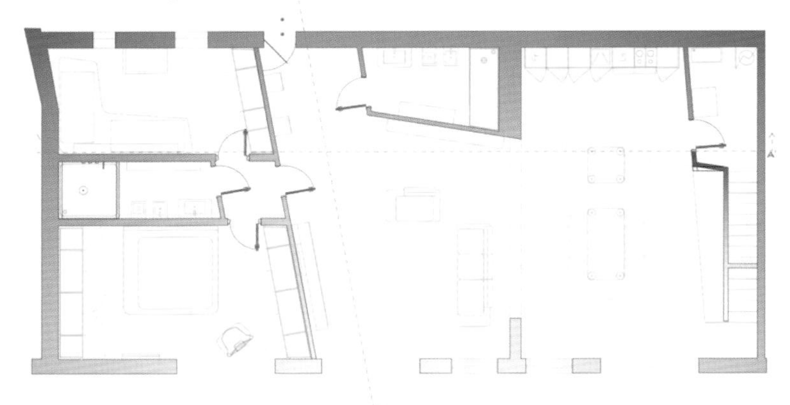

shower shower shower shower shower

to
dream

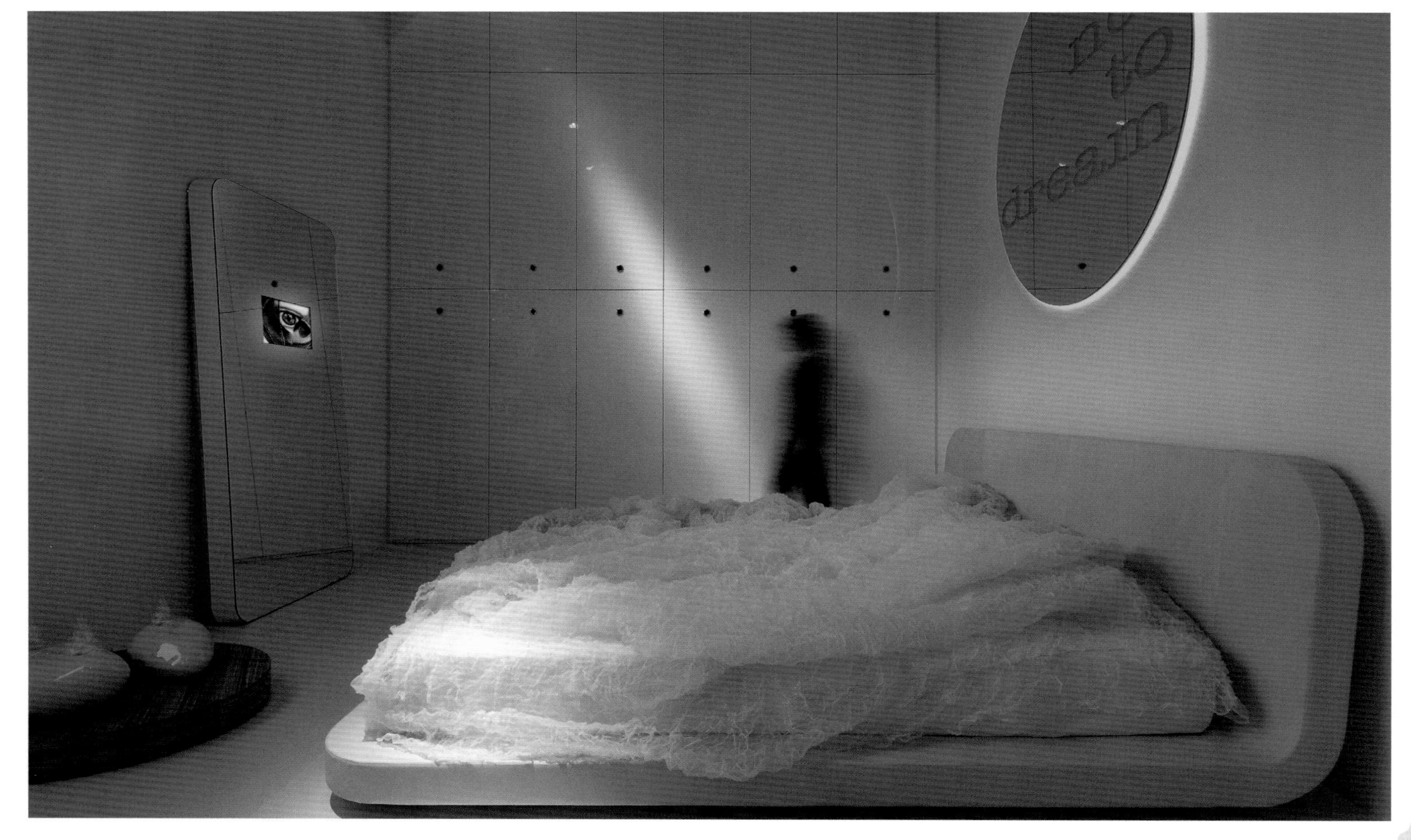
to
dream

Project Name: Annie House
Design Company: Bercy Chen Studio
Photographer: Mike Osborne

项目名称：安妮别苑
设计公司：贝震西设计工作室
摄影师：麦克·奥斯本

红与蓝的浓情华尔兹
WALTZ

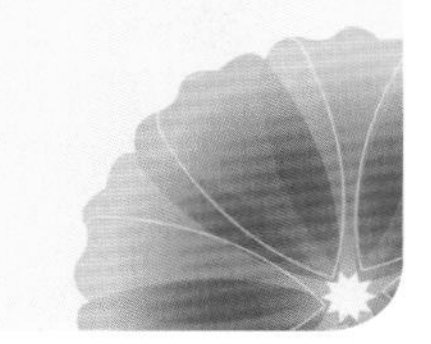

本案基地狭小，为填充地块。两独立生活区域，外似亭台，以玻璃走廊相连，极大地方便两户共居。
地块中央，一小池碧波涟漪，绿色园景。两亭台如玉盘中落，抒中国江南园林之意象。一亭台内设两卧、一浴；另一亭台则容其他生活空间。
小池涟漪聚焦空间视线。量体绕以围墙，如同天井之平面。前是平院平台。后院、户外的生活起居空间，内驾一半透明玻璃小桥，彰显生活之精致。
建筑模块钢构，预制板合围，乃是环保、高效的实践。其一亭台，二楼桁架设计，如同桥梁，工程力学的合理运用，最大限度地减少一楼承受的垂直压力。
平面屋顶，是另一个户外的天地，绿植中成就着一个别致的露天茶座。可伸缩遮阳篷功能多重，既可作孩子玩耍的场所，又可作航海物件的收纳空间。
如果说建筑是文化的传承，本案则是多元文化的体现：摩尔式建筑的室外空间、遮阳设备；亚洲建筑的水景设计，空间延续；日式亭台楼阁的透明结构，精简内饰。

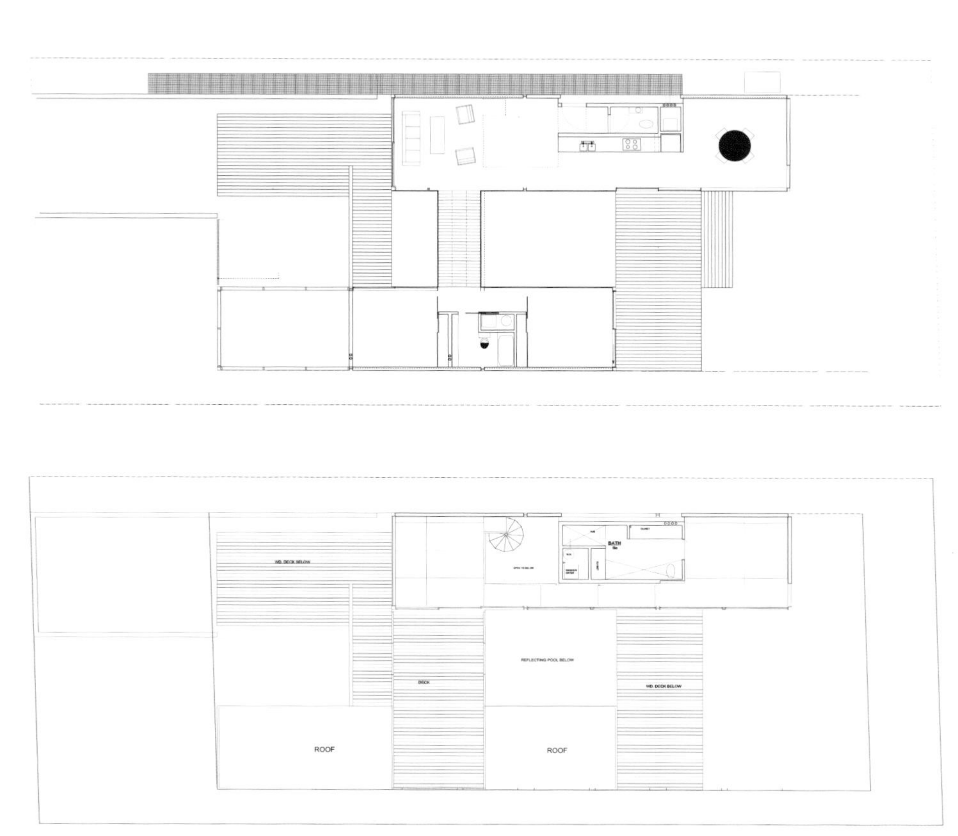
WD. DECK BELOW
BATH
REFLECTING POOL BELOW
DECK
WD. DECK BELOW
ROOF
ROOF

The house is located in south Austin on a small infill lot, built for two families and therefore split into two living areas, and consisting of two pavilions connected by a glass hallway.
One pavilion contains two bedrooms and one bath while the other contains the rest of the program. Each volume is placed against the side setback of the property creating a central water garden in-between.
The reflecting pool becomes the focal point and all sides of the house open onto it. The walls against the sides of the property are closed, creating a courtyard layout. The two parts of the house are staggered to create a deck area in the front as well as a more private outdoor living area in the back, visually united by the translucent glass bridge.
The house is constructed of a modular steel frame, infilled with prefab thermasteel panels to minimize construction on site waste. the structural frame is exposed, showing the construction process and articulating the house' s facades. the repetitive modular method as well as the prefabrication allows for greater efficiency during construction. The 2nd floor in one of the pavilions is a viereendeel truss which acts like a bridge and minimizes the number of vertical structural supports in the 1st floor.
The flat roofs allow for terrace spaces which creates additional outdoor areas for plants and alfresco dining. The roof space is covered with a retractable awning made of shading tarp for nurseries and hardware from the nautical industry.
The house is influenced by different regions and cultures. both the use of the roof as an outdoor living space and the shading devices are derived from moorish architecture. The body of water and the spatial continuity between inside and outside was inspired by Asian architecture. The structural transparency of the volumes and the minimalist aspect of the interior was derived from japanese pavilions.

Project Name:Celio Apartment
Location: Rome, Italy
Design Firm: CarolaVannini Architecture
Photographer: Stefano Pedretti
Surface: 240 m²(interior space) + 20 m² (exterior space)

项目名称：Celio 公寓
项目地点：意大利罗马
设计公司：CarolaVannini Architecture
摄影师：Stefano Pedretti
项目面积：240 m²（室内）+20 m²（室外）

精于内，简于形
REFINED INSIDE, SIMPLE OUTSIDE

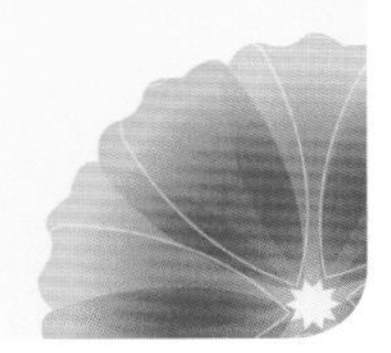

这所靠近 Colosseo 地区的豪华公寓经过设计师的重新设计，满足了户主连通室内外景观的需求，梦幻的房间与灯光更是此公寓的亮点。建筑师在屋内增开了几个窗户，室内的格局也作出了相应的调整，从而制造出多重的视角享受。

公寓主要的连结空间是厨房，连通公寓的入口和起居室。透过连续的 LED 灯光，建筑师把厨房分割于天花和地板之间镶嵌的墙壁，另外，两扇滑动玻璃门使得厨房可随意开合。

起居室极具简约之风，而其他由建筑师设计的家具更是与室内的简约风格相得益彰。为了强调公寓的视角与空间深度，建筑师细心地研究公寓的灯光系统，同时室内延伸至阳台的木地板无意中也成为连接室内外的桥梁。黑色的门窗框仿佛把室外的自然景观收进屋内，营造一种置身油画的感觉。

公寓的休息区域包括三个睡房，三个浴室和一个办公空间。休息区域由一道走廊连接，建筑师通过对灯光和特殊位置的运用，开阔了走廊的空间。走廊墙上一排长长的内嵌柜橱完美地与室内的建筑融合。在这里建筑师放弃了使用传统的柜橱门，转而使用树脂玻璃面板，并在柜橱里面加上灯光。

主人房通过两个大窗户自然采光，房内还设有一片由灰色树脂覆盖的美妙休闲区。灰色树脂延续了灰色木地板的作用。休闲区的平台上还摆放着一座极可意浴缸，两

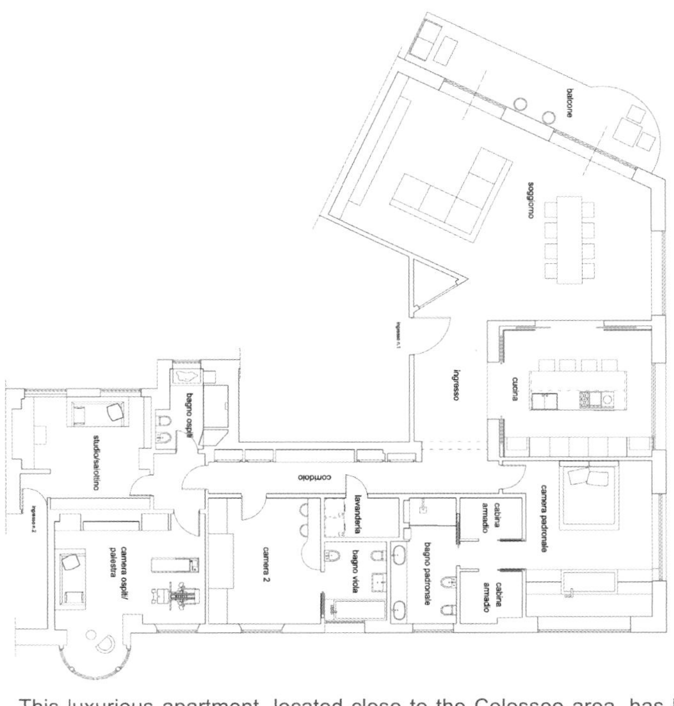

道平行的的墙壁通往大衣橱，大衣橱后面便是主浴室。

主浴室色调和主人房一致。在灰色墙壁的衬托下，主浴室内白色的更衣室与陈列柜显得尤为突出。四个绿色的木质陈列柜是浴室内唯一的色彩元素。

工作室位于走廊的尽头。它有自己独立的门口，方便户主灵活独立地运用工作室。其内温暖而朴素的木家具全由设计师设计。

客房（也作健身房）浴室采用大红色调，在内可以舒服地泡澡，也可以正常地淋浴。这是主人休憩、锻炼和打发时间的空间。

儿童房和其专属浴室有别于公寓的其他空间，紫色是儿童房及其浴室的主色调，这里是一个童话的世界。房间内的一切都令人忆起儿时的玩乐时光。

This luxurious apartment, located close to the Colosseo area, has been reorganized through a complete and detailed design project.

The client's main need of opening the space toward the outside landscape, generated a design characterized by light and airy rooms.

Several windows have been reopened, and the interior distribution has been changed in order to create multiple perspectives.

Main space of the day area is the kitchen volume, which is in direct relation with both, the entrance and the living room.It is designed as an isolated volume, separated from ceiling and floor,through continue led lights. Two

sliding white glassdoor sallow to open and close the kitchen, depending on the user' s needs.

The living room has a minimal flair, and its furniture (designed by the architect) creates a balance with the interior architecture.

The lighting system has been carefully studied in order to emphasize perspectives and space depht.

A direct relation with the exterior area is underlined by the wooden floor that exetends from the interior space into the balcony.

The blackdoors and windows frames, create a painting-like effect by framing the surrounding natural landscape.

The night area has three bedrooms, three bathrooms and one office space.

The corridor that leads to them,has been designed in order to open up the space through the use of niches and lights.

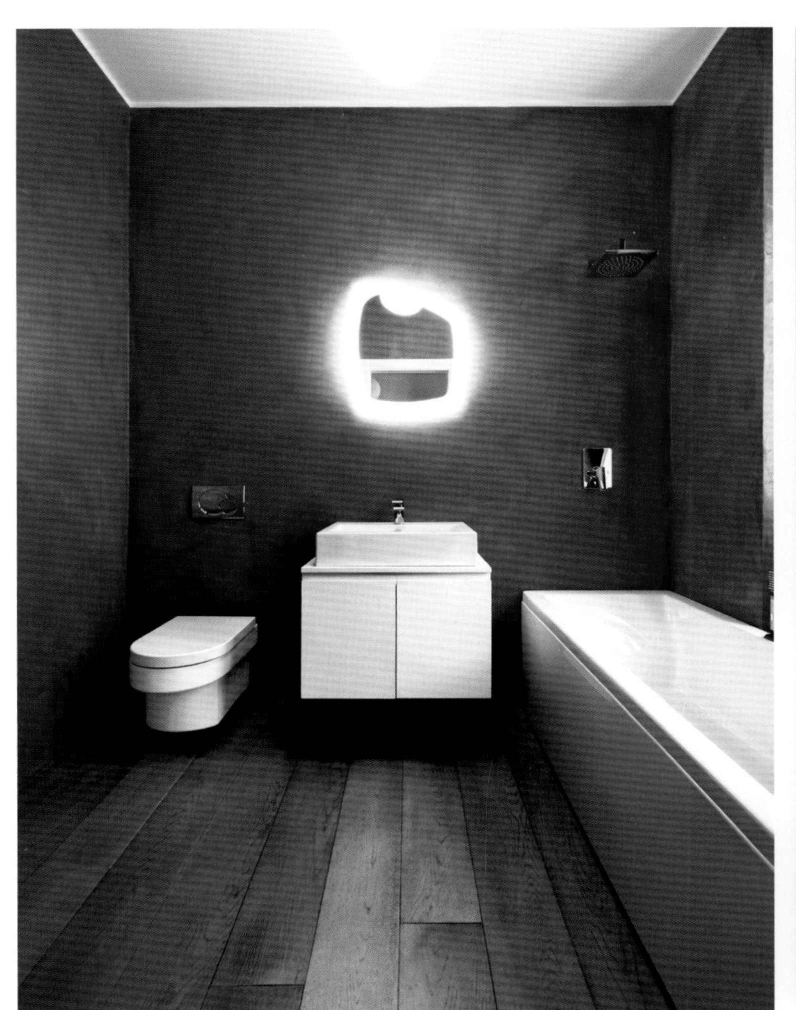

A long built-incupboard, which perfectly merges with the architecture, gives up on classical doors and replaces them with backlit printed plexiglass panels.

The master bedroom is naturally lit by two big windows and has a nice banked relax area,coated with grey resin. This material creates a continuity with the grey wooden floor.

A jacuzzi is located into the banked platform.

Two simmetrical walls lead into the walk-inclosets area and, then, into the master bathroom.

The master bathroom has the same bedroom' s colors and geometries.

The white bathroom fittings and cabinets, stand out against the greywalls and four green wooden cabinets are the only colored elements.

The kid' s bedroom with its coordinated bathroom,creates a fairytail world,separated from the rest of the apartment.

Everything recalls play times. Purpleis the main colour in both the bedroom and bathroom.

The bathroom' s fittings are emphasized through the purpleres in walls veneer.

In the guestbedroom (also used as a gym area)relax, exercise and free time spaces alternate.

The guest bathroom is characterized by a strong red color. It is located close to the gym and has a confortable hammam space,also used as a normal shower.

The office is placed at the end of the corridor and has its own entrance,which allows a flexible and indipendent use of it.

The wengé wooden furniture, with its warm and sober colors, is designed by the architect.

Project Name: The Peninsula Residence
Design Company: Bercy Chen Studio

项目名称：半岛别墅
设计公司：贝西震设计工作室

半岛雅居，动静相宜，缤纷相伴
THE PENINSULA HOUSE

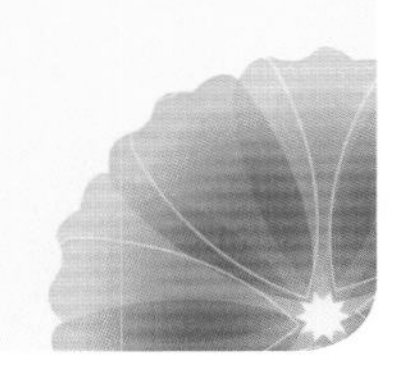

半岛别墅，内置家具虽传统，却不乏高端欧式品牌，同时也有诸多定制家具来界定各个空间。

二楼书房，组合学习、游戏桌柜，赋予该处空间整体木雕之感。轻松、实用，极大地方便书房独享及宾客来往。男孩房的内置床铺，如同豪华游艇的船舱及列车的高级铺位。设计灵活、方便，一旦孩子成人，不需大费周章，即可华丽转为成人房。楼上视听空间，组合弧形沙发、坐榻，轻轻打开，成就着另外一个卧室。单一的空间却有着复合的功能，这是本案设计的一大特色。如同卧室，定制的折叠门，开启时，外面是工作、收纳、娱乐的空间。闭合时，自是一室清静，一室祥和。工作上需要时，更可成为办公的中心所在。

起居室内，谈话空间，气氛亲密。平台陡然而出，眼前，碧波万顷；身后，家居安怡。

The Peninsula Residence strikes a balance between built-in and conventional furniture. Although the interior design relies on high-end European furniture brands, including Flexform and Team 7, the numerous integrated furniture elements define several spaces.

In the second floor library, integrated desk and game table amplify the illusion that the library was carved from a block of wood. Both playful and functional, the library facilities private contemplation and social interaction. The built-in bunks in the boy's bedroom remind one of the cabin of a fine sailing yacht or the berths on a luxury train. The room is the perfect escape for a younger child, but is designed to convert to a full size bed when the owner's child reaches the teenage years. The integrated curved sofa/lounger in the upstairs television room provides a comfortable setting for watching a movie and doubles as a 5th or 6th bedroom.

The project asks many rooms to serve multiple functions. In the master bedroom, custom folding doors conceal an efficient work space, storage and television. When closed, the master bedroom stands as serene, restful retreat. However, as needed, the functional aspects of the work center can be accessed.

The conversational area in the family room creates an intimate social space. The raised platform provides a shift of perspective that changes the encounter with the home and the lake.

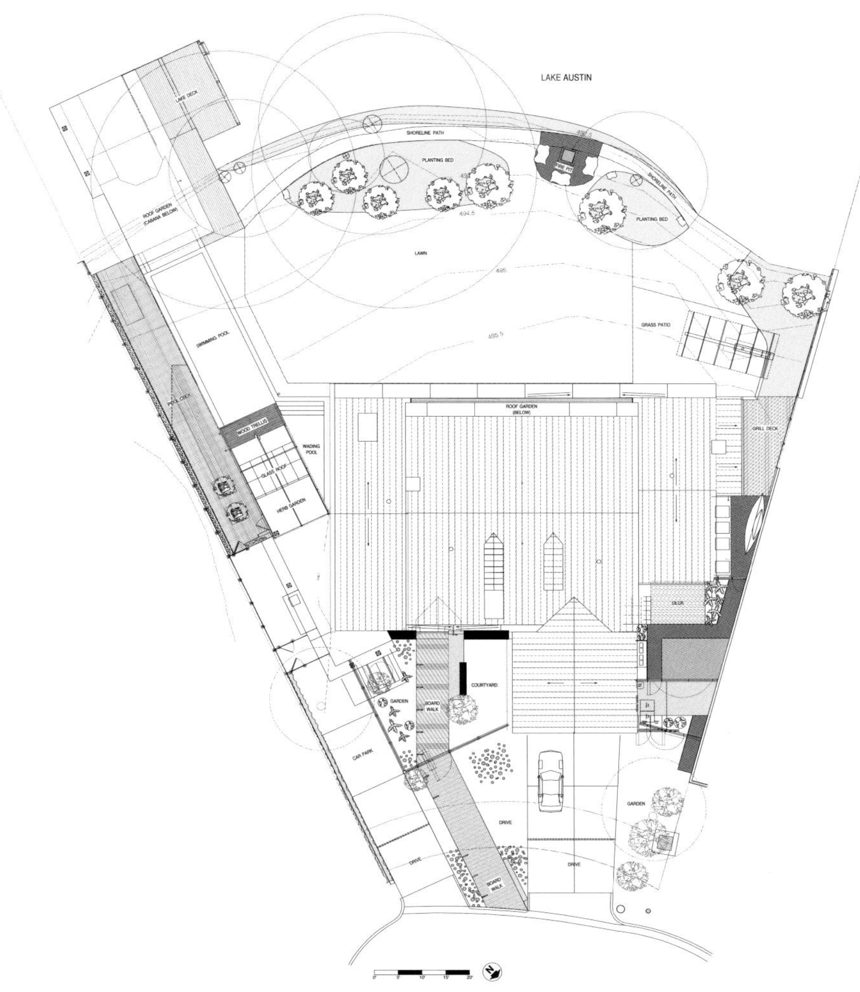

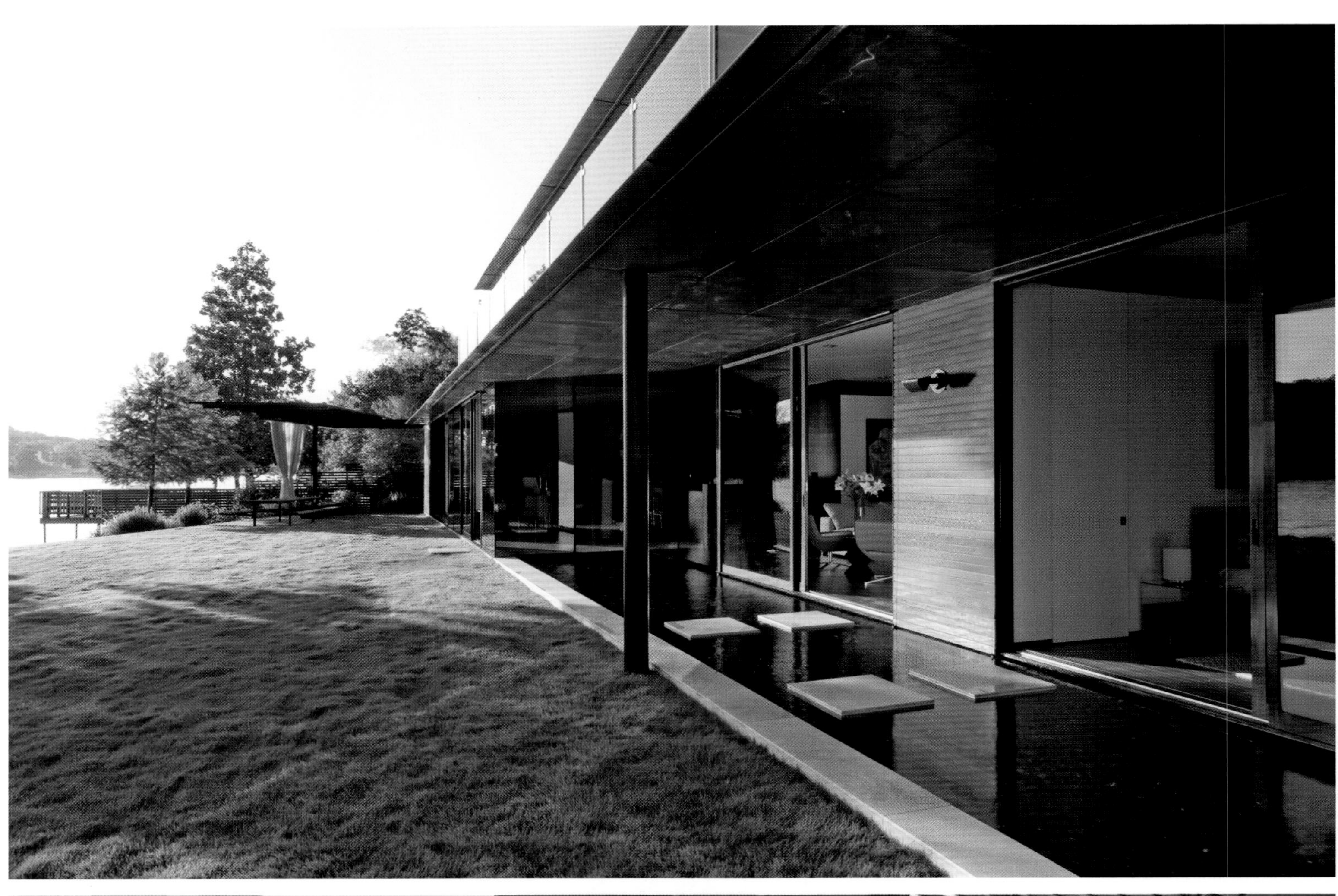

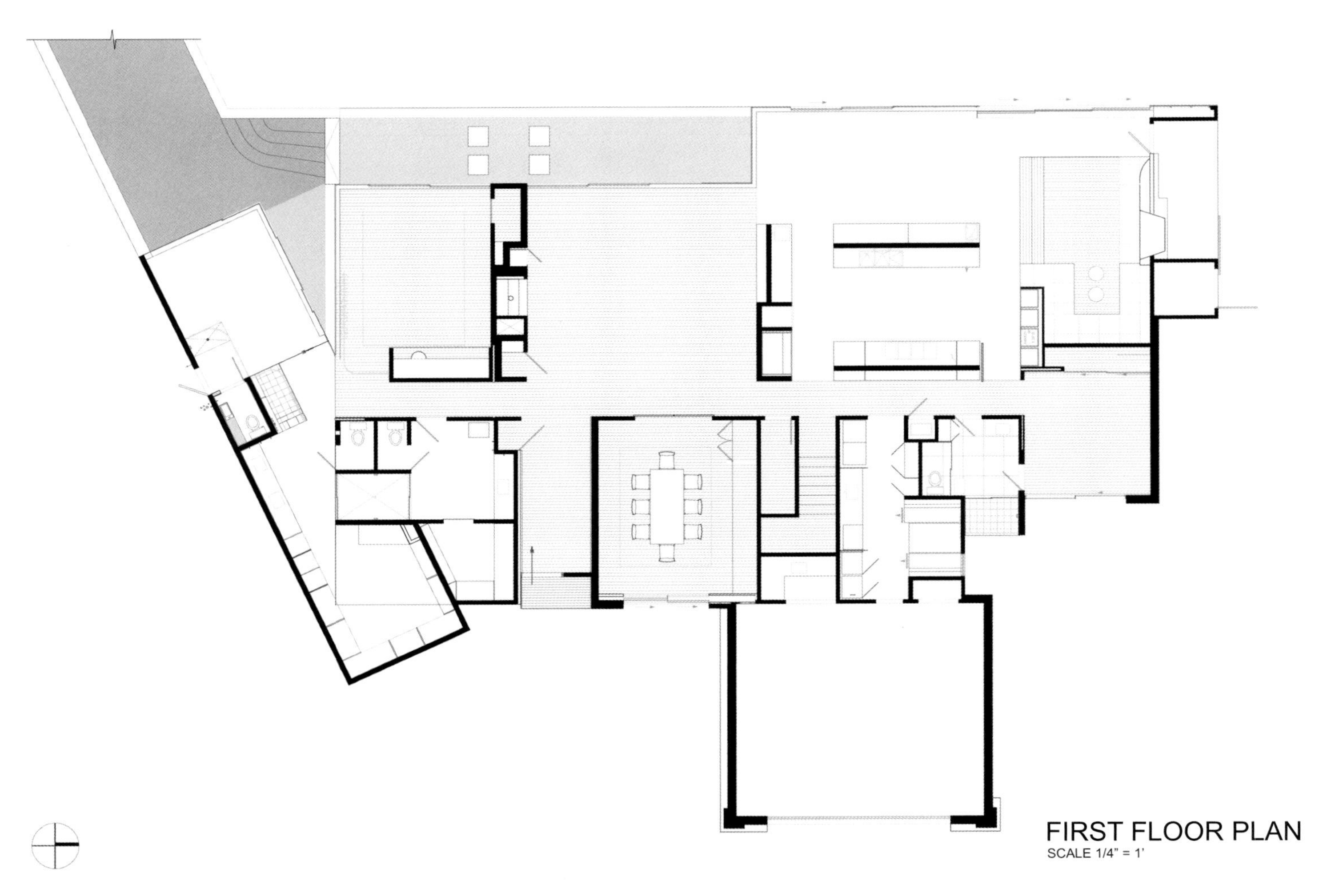

FIRST FLOOR PLAN
SCALE 1/4" = 1'

SECOND FLOOR PLAN
SCALE 1/4" = 1'

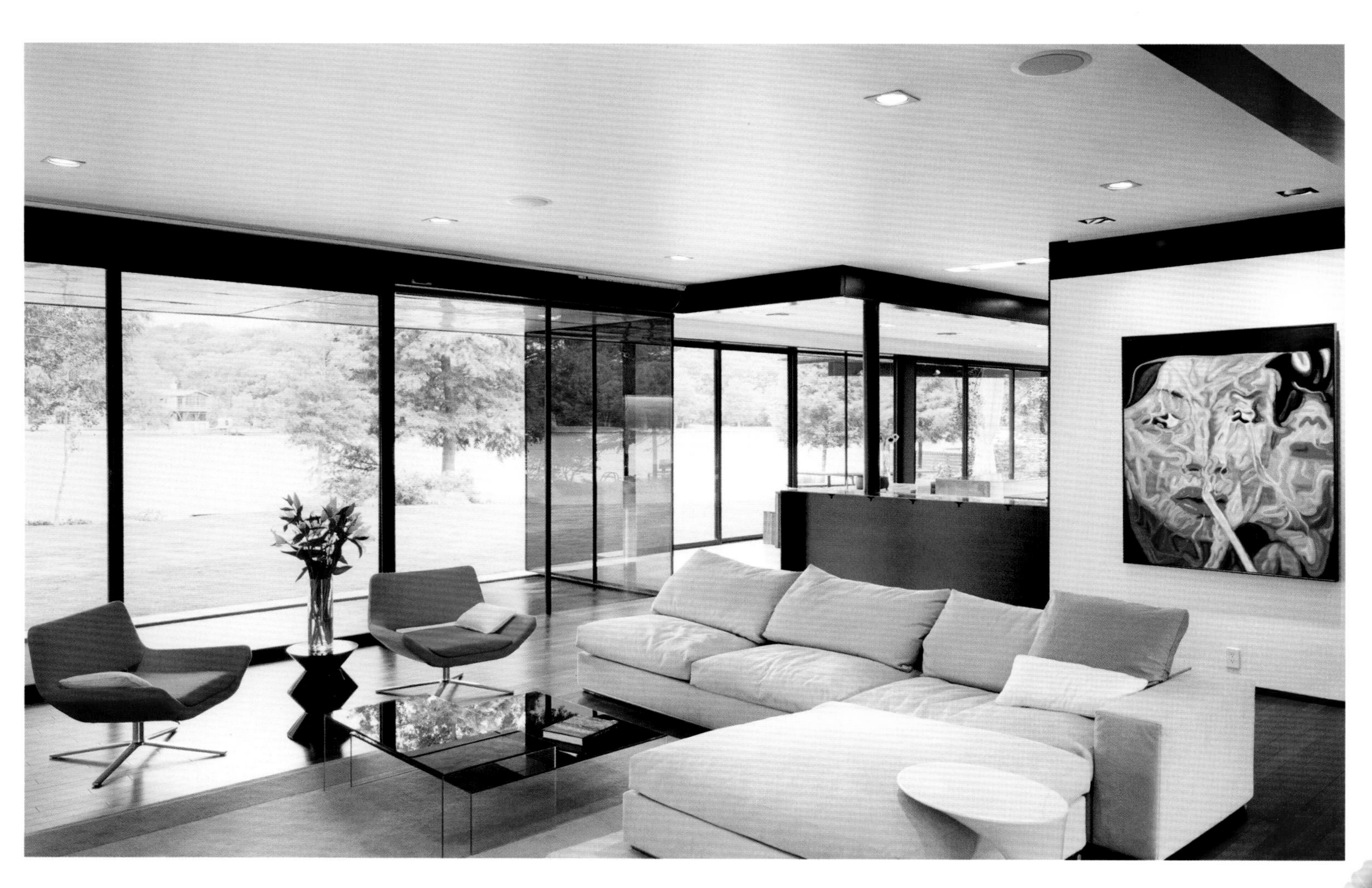

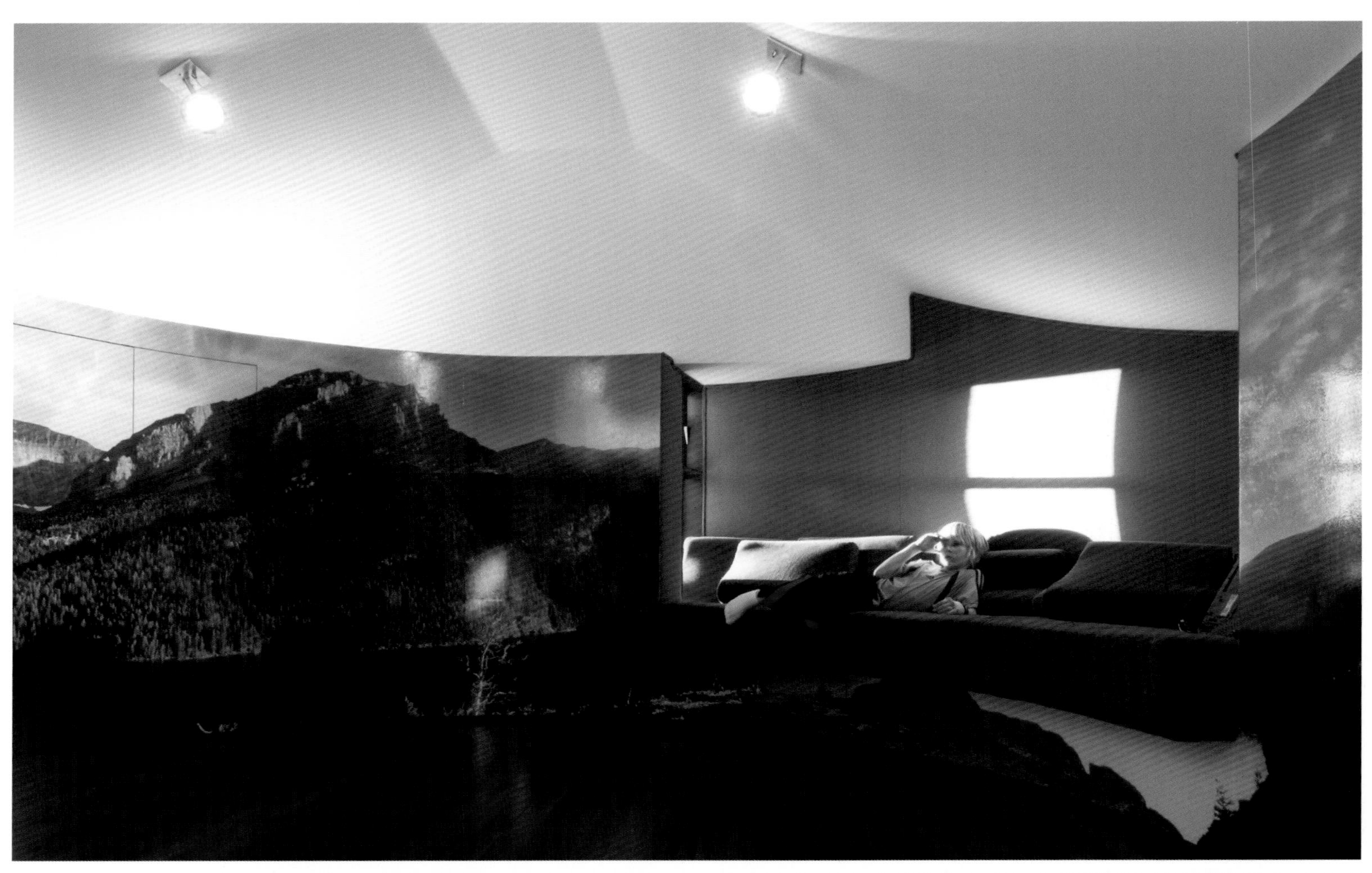

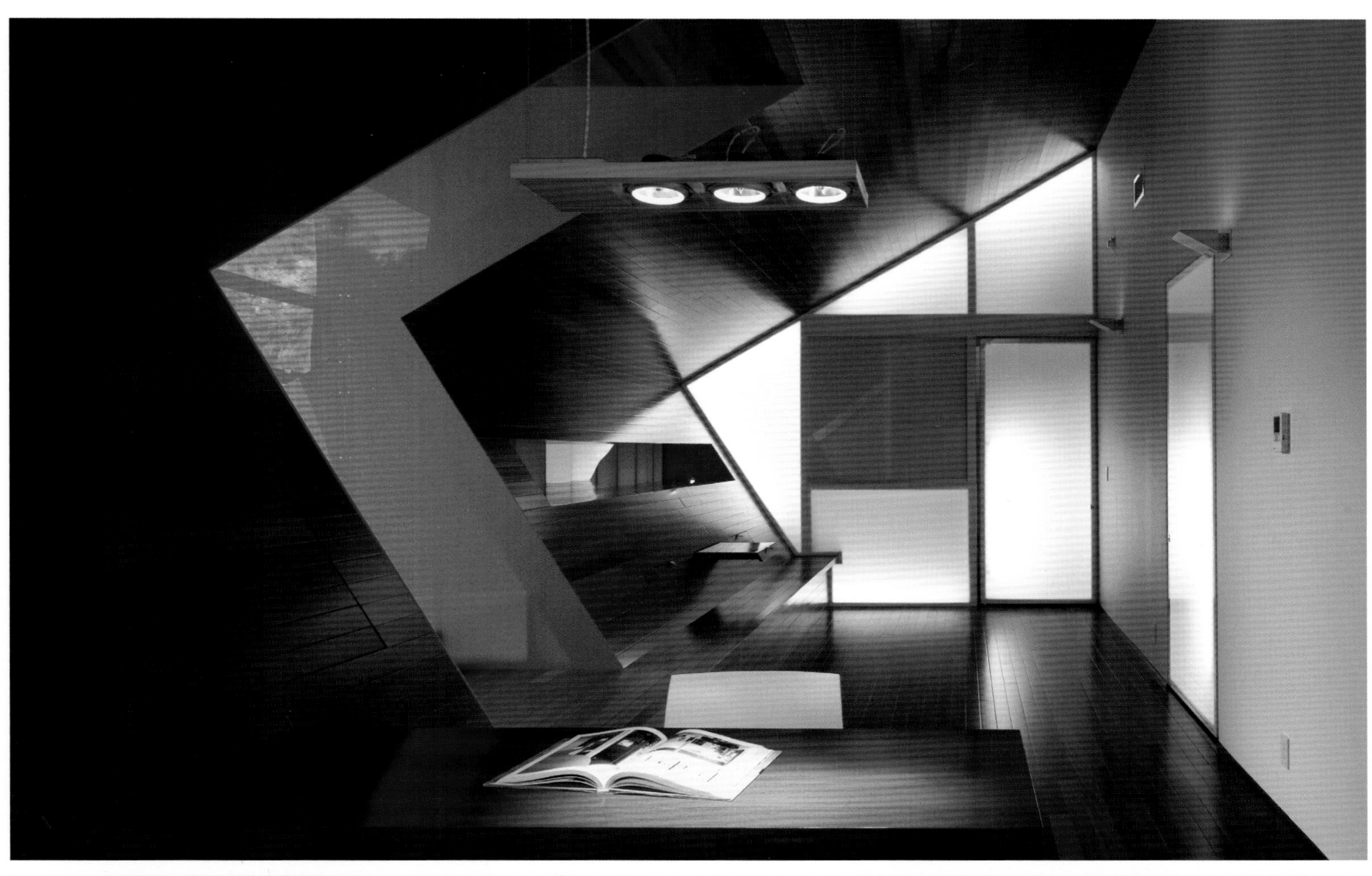

Project Name: Venice Bungalow House
Design Company: Jamie Bush & Co.

项目名称：威尼斯平层别墅
设计公司：杰米布什室内设计

用艺术照亮生活的每个角落
THE SPACE LUMINOUS GLOWING

如果空间用”珠宝盒”来比拟，则空间很有可能变得逼沓、拥挤，但本案除外。本案是原建于上世纪 50 年代之量体，空间有限。借设计师布什之妙手，不足 112 m^2 的空间，却尽拥珠宝盒之气质，彰显空间盈动、开阔之势。

狭小空间，大手笔铺陈铸就。加尔各答产的金色、灰色大理石出卧室，入厨卫，到生活起居区，一路蔓延，到达户外庭院。各功能空间、室内、室外因此合而为一。视觉在空间中可以纵横伸展，视线通透，一览无余。静读于书房之内，可以放眼音乐室、起居室、卧室之间。家居生活之温馨、快活、舒适之感尽在本案空间体现。各功能空间，天花用品牌墙纸书写温柔气质，彰显夺目。卧室是谷物纹样，音乐室、书房则是植物纹样。除墙纸大量使用外，另有瓷砖、织物、木材不失时机加以点缀。厨房沙黄色陶瓷砖、书房漂白橡木地板，其中些许壁板，同样质地，内置镶嵌，打造“日本桑拿浴箱效果”。卧室里，几束淡色薰衣草，比利时亚麻布苇打造温柔的气象。还有墙体、定制的混合品牌漆粉刷，与布苇呈合围，护卫之势，更显空间的气派与恢宏。

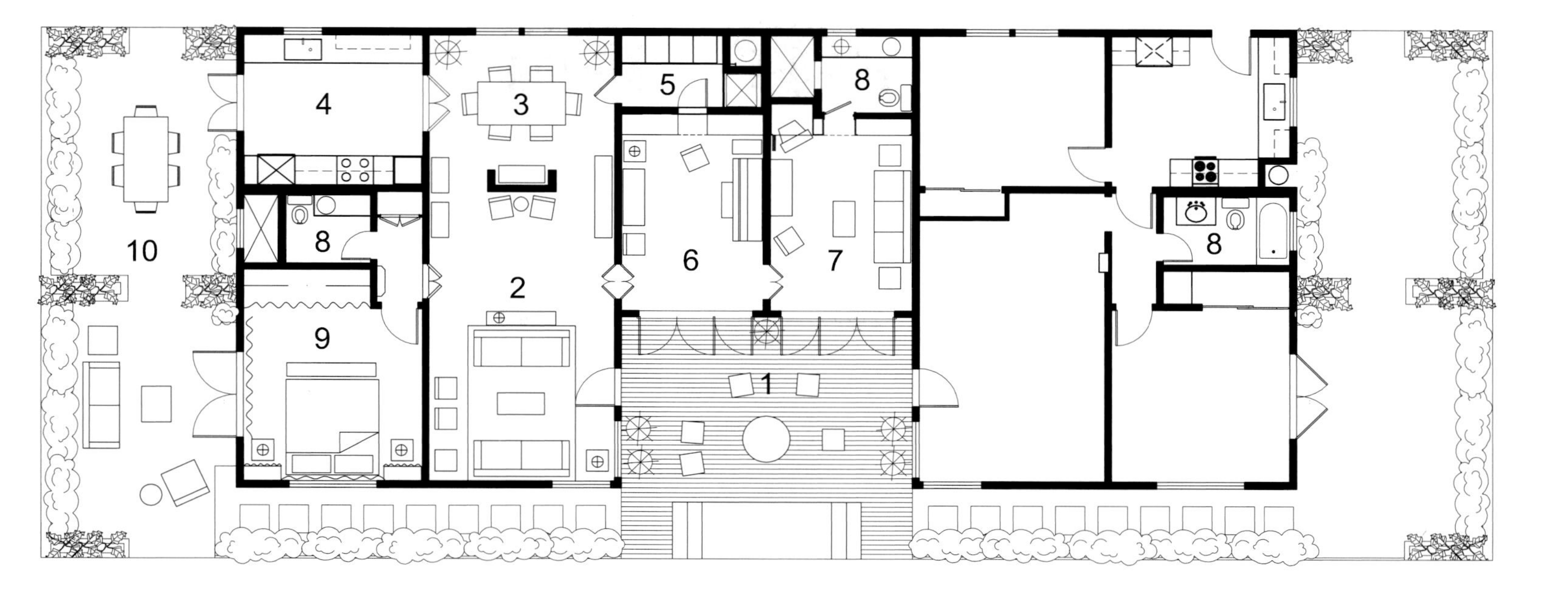

The term jewel box may be an over used one. But when it comes to Jamie Bush' s dazzling dot of a Venice Beach House, the expression is more than justified. It is here, in a series of petite rooms with scant square footage (all told, less than 1 200), that the designer' s talent gets huge room to roam. By blurring interior and exterior areas, making premeditated plays on scale, and adding just-so doses of color and texture, Bush pulls off a real-deal decorating feat: He doesn' t just make small feel bugger. He manages to make a diminutive 1950s box feel downright opulent.

"The key to small spaces is that you have to make room for a few grand gestures," says the designer, who has a master' s degree in architecture form Tulane University and has worked for the likes of Kelly Wearstler and architects Marmol-Radziner. In the case of his own home-which he shares with Stephen Calipari, a piano teacher- one of the grandest of all gestures includes his use of Calcutta gold marble. The gray and gold veined white stone extends from bedroom to kitchen to living area-even to the outdoor patio.

Both the width and depth of the space are broadened by unfettered sightlines.

PERSIA BRIDGE OF TURQUOISE
THE ORCHID
SHELLS
X-RAY
Creature

Stand in his 10 foot by 15 foot library, with its 9 foot long Futurama in Los Angeles sofa and gallery size abstract painting, and you can see all the way through the music room and the living room to the bedroom hallway.

Bush also draws the eye up by wallpapering many of the ceilings. In the bedroom, he uses a grain-like Keith McCoy paper on the ceiling. In the music room and library, it is a custom McCOY fern pattern. He also "Wraps" rooms in tile, fabric and wood. In the kitchen, the sand-colored ceramic square penny tile isn' t just used as the expected backsplash. Instead, Bush went whole hog. I" made a rule," he says. "If you are a vertical surface in the kitchen, you get covered in the tile. It is a more urbane gesture." And for the library, his favorite

room in the house, he took the bleached white oak he used for the floors and extended it up onto the walls and built-ins for a "Japanese sauna box effect." Similarly, for the bedroom, Bush took a few bolts of "faded lavender" Belgian linen and draped it along almost every vertical surface, effectively covering every square inch of wall in yard after yard of ripple-fold fabric. The lone exposed wall is painted in custom-blended Pratt & Jambert that matches seamlessly with the drapery. "Too many small moves that are all the same scale make rooms feel fussy," say Bush. " And in the bedroom, it rally fells enclosed by the color, and that ends up feeling grand."

Project Name: Glenbrook Residence
Design Company: David Jameson Architect
Site: 4 206 m^2
Building: 1 050 m^2

项目名称：天井大公馆
设计公司：大卫•杰姆逊建筑师事务所
基地：4 206 m^2
建筑：1 050 m^2

栖居山林，展翅欲飞
FLYING ABOVE HILLS AND FORESTS

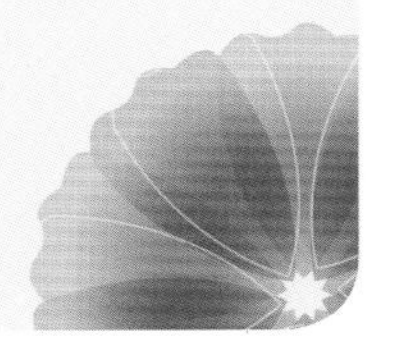

本案正如其中文名称，两面厚重的墙体，围合一建筑量体。绿树丛林，概念建筑手笔，描绘公、私两处亦或独立、亦或合二为一的空间。两者之间，另有房舍片片，连接过渡两处空间。

公、私各翼，设计如同地球之形体，以其厚重、静态之材质、之形体，相互连接，共同支撑着建筑量体、檐篷、亭舍。

生活馆舍水晶般闪亮，自然建筑的中央，内设烹饪、饮食、生活等功能空间。各翼之上，另有轻盈折叠华盖。六面整体设计，片片独向山林，好一个月朗风清的去处。华盖之下，玻璃幕墙，创浮顶之意象，模糊着划分明显的内、外界限。各元素棱角分明，尤其是动态设计，显示着空间的流体质感。

Shaped largely by the site, the Glenbrook Residence is conceptually a courtyard inserted between two heavy walls. Threading the walls through the treescape to create distinct yet connected structures allows the house to be divided spatially into the most public, most private and a living pavilion that can become either or both. The residual in-between spaces create outdoor rooms that engage the building.

The public and private wings of the house make up the foundations of the design concept. They are thought of as being of the earth and are articulated through their materials and shape as heavy, static pieces. These wings define the bounds of the house and act as the backbone to support the various courtyards, upper roof canopies and the dynamic living pavilion that sits between.

The living pavilion is conceived as the center-piece of the concept and glows like a crystal between the heavy wall elements of the house and contains the cooking, eating, and living spaces.

Above each of the heavy wings floats a thin, folding roof canopy. More than a simple surface, this roof canopy is conceived as an entity where nothing is hidden and all six sides are exposed to view. The walls that contain the spaces beneath these canopies are made of glass to create the illusion of a floating roof and to blur the boundary between inside and outside. All of these elements adopt a language of angular, dynamic forms in order to be completely liberated from the solid elements of the house.

Project Name: Beverly Skyline Residence
Design Company: Bercy Chen Studio

项目名称：贝弗利天际大公馆
设计公司：贝西震设计工作室

晨曦暮霭，坐享林间观景台
THE FORESTLAND HOUSE IN DAWN AND DARK

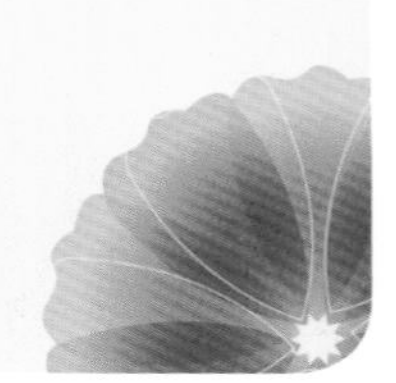

德克萨斯州奥斯汀市，贝弗利天际大公馆，有着一英亩宽大基础，260 m^2 的空间，单栋独享。

原本只打算小改动的个案，借本案设计师之手，重新焕发生命活力。室内外华丽转身，新添房舍，园内新近开掘，园景新布置。设计手法纷呈多样。雨水收集池、蓄水池、反光小池、可回收建材彰显绿色、环保设计。布局的改变，涉及面广，但却有效地理顺了建筑、地形及风景之间的关系。

基地边缘，花园静立，小河流淌，如何使量体与之连接，成了本案的一个重要目标。于是，设计师施以妙手，在量体周围绕以雕栏、玉砌，凭栏四望，秀木毓水，尽入眼底。此设计灵感源于日本京都的清水寺，该案建于公元 7 世纪。

暮鼓晨钟，城市风光独好。宽景平台，令人不禁感叹人与风景竟然如此接近，如此和谐，如此轻松。

空间内外，极尽可再生材料之使用。池塘、蓄水池的水资源利用，不仅是对资源的有效使用，更是建筑与自然进行的又一种对话。

铺陈，艺术、雕塑质感，提升空间气韵，彰显其内涵。

花园植物，土生土长。树林、灌木，成年植株，遍布各处，展示着当地特有的风情。不仅简约、不仅实用，本案更是空间秀美的化身。这让业主放心，更使得设计师舒心。四十多年，风雨穿越，空间是时尚的展现，是对绿色技术的有效使用，更是一种持久，正欲走过经久的风雨。但，现代依然。

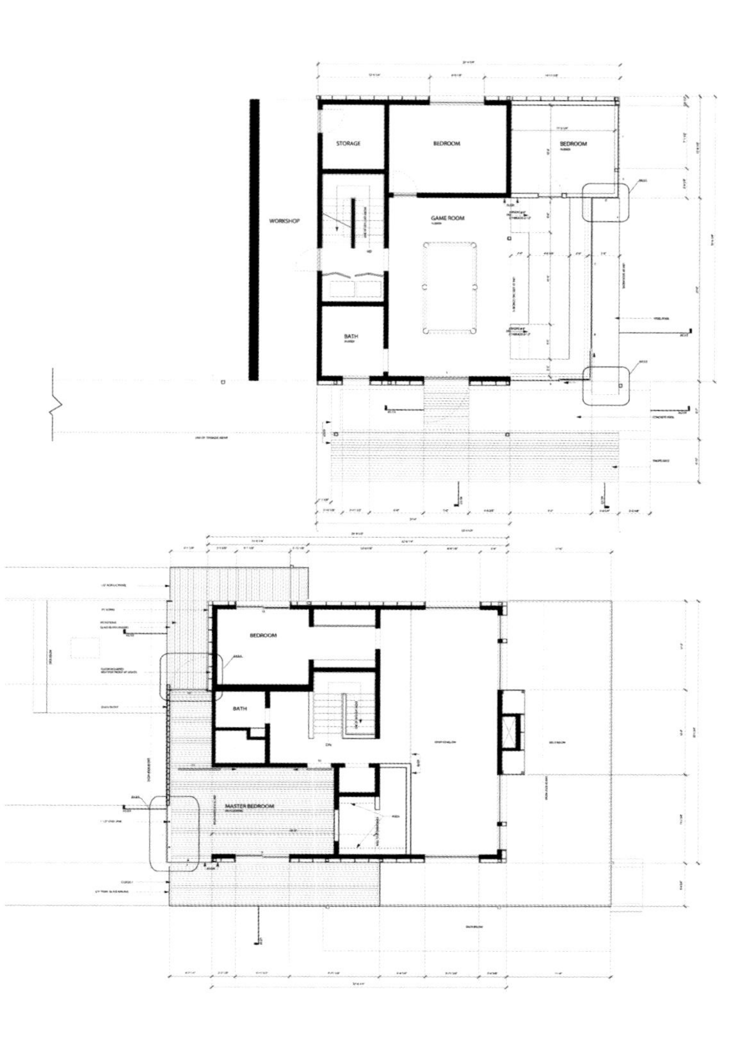

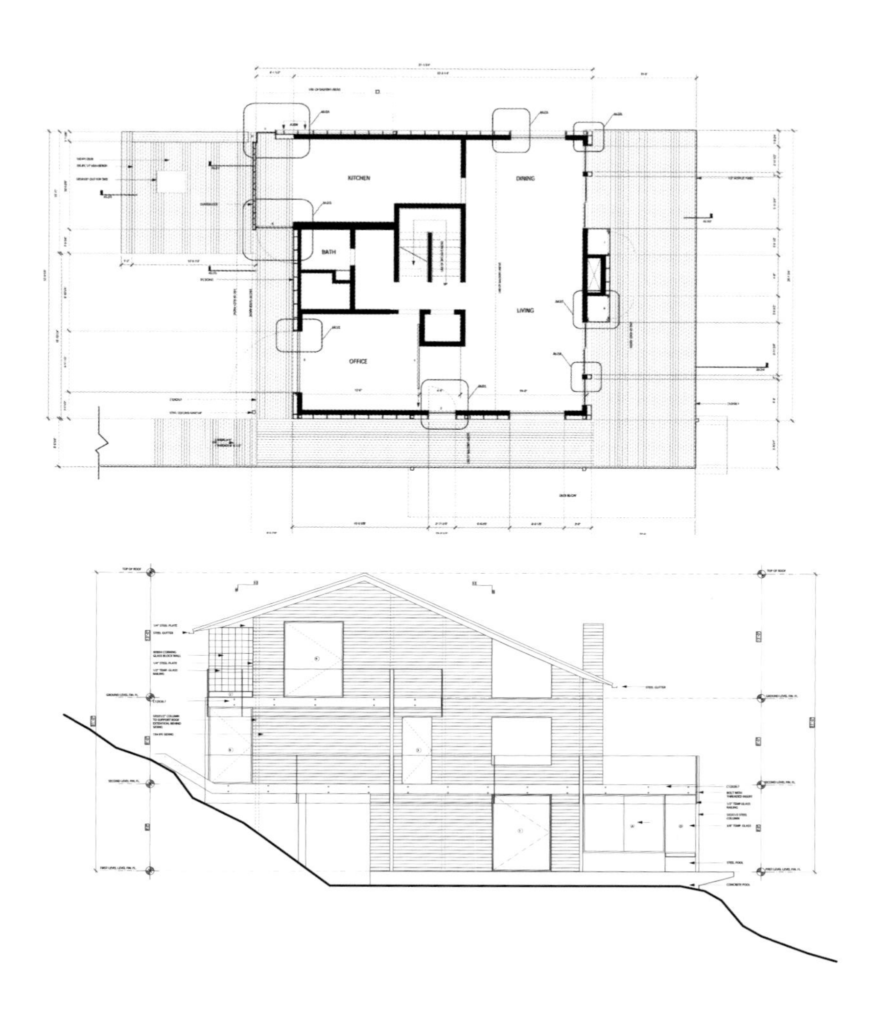

Beverly Skyline Residence, located in Austin, Texas, is a single-family residence and consumes 2 800 sq. feet of its 1-acre property lot.

Though the original intent for the home was to be a standard, modest remodel, Bercy Chen Studio took the design one step farther and transformed the project into a full site plan remodel. The house was transformed into a complete interior and exterior remodel with a new room addition and included an excavation and re-landscaping of the garden and parts of the yard. In addition, the house was equipped with recyclable, green technologies such as rainwater catch basins, cisterns and reflecting pools, and the use of recyclable building materials. As the original house was poorly sited, a large motivation for the remodel was to reconnect the house with the sites steep topography and capture the expansive views.

One goal was to integrate the architecture with the native garden and creek at the bottom of the property. The inspiration came from the Kiyomizu Temple in Kyoto, Japan (founded 7th century A.D.), which sits above the landscape and provides panoramic views of the city. Similarly, in order to fully enjoy and reclaim the home' s views, the house is wrapped by exterior decks and glass railings, which provide an uninterrupted view of the valley and woods beyond. These seemingly borderless edges reconnect the home with the nature of the site. Also, by maximizing the amount of exterior deck space, people can experience the views in a closer and more personal way.

Recycled materials were employed at every possible opportunity. The project also makes extensive use of harvesting rainwater stored in pools and reservoirs to re-connect the house with its site.

Architecture elements are treated as installation artwork and sculptures in this project.

The selection of plants in the garden is primarily plants native to the central Texas region. The use of native vegetation contributes to the sustainability of Beverly Skyline by minimizing the usage of water by following the overall principles of xeriscaping. The garden is planned around existing mature trees and shrubs with various ground covers and perennials. The intention was to preserve the characteristics of the site as much as possible and retain the essence of a landscape native to the Edward' s Plateau in the hill country.

The appeal of the redesigned house is not just its simplicity and functionality but also its well thought out aesthetics. Overall, both the client and architect were happy with the finished remodel. Beverly Skyline Residence is no longer an outdated 1970s residence, but is now a sleek modern home equipped with green technologies and better energy efficiencies that will last for years to come.

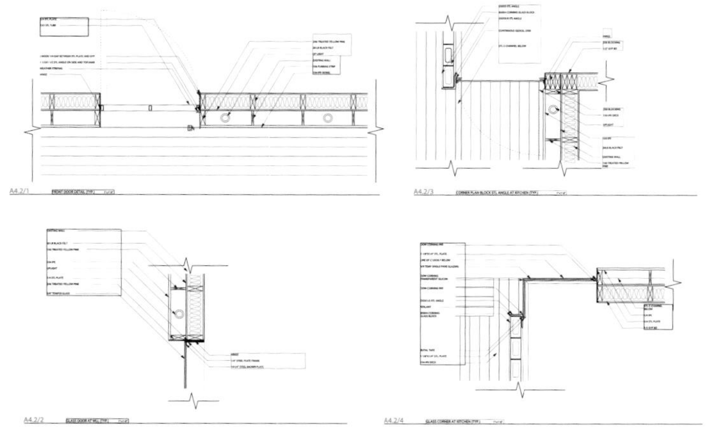

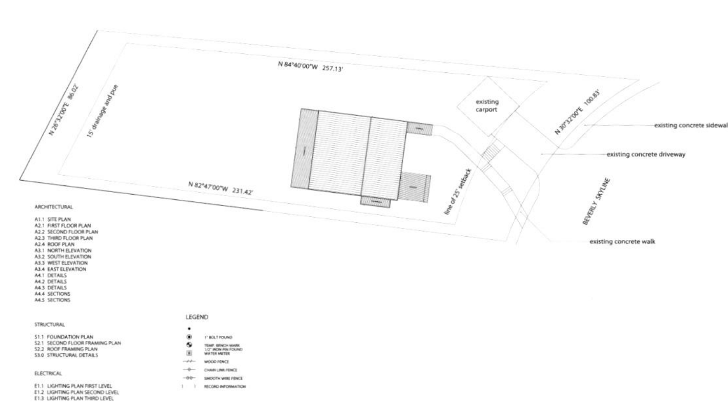
N 84°40'00"W 257.13'
N 26°32'00"E 86.02'
15' drainage and pue
existing carport
N 30°32'00"E 100.83'
existing concrete sidewalk
existing concrete driveway
N 82°47'00"W 231.42'
line of 25' setback
BEVERLY SKYLINE
existing concrete walk
ARCHITECTURAL
A1.1 SITE PLAN
A2.1 FIRST FLOOR PLAN
A2.2 SECOND FLOOR PLAN
A2.3 THIRD FLOOR PLAN
A2.4 ROOF PLAN
A3.1 NORTH ELEVATION
A3.2 SOUTH ELEVATION
A3.3 WEST ELEVATION
A3.4 EAST ELEVATION
A4.1 DETAILS
A4.2 DETAILS
A4.3 DETAILS
A4.4 SECTIONS
A4.5 SECTIONS
STRUCTURAL
S1.1 FOUNDATION PLAN
S2.1 SECOND FLOOR FRAMING PLAN
S2.2 ROOF FRAMING PLAN
S3.0 STRUCTURAL DETAILS
ELECTRICAL
E1.1 LIGHTING PLAN FIRST LEVEL
E1.2 LIGHTING PLAN SECOND LEVEL
E1.3 LIGHTING PLAN THIRD LEVEL
LEGEND
1" BOLT FOUND
WOOD FENCE
CHAIN LINK FENCE
SMOOTH WIRE FENCE
RECORD INFORMATION

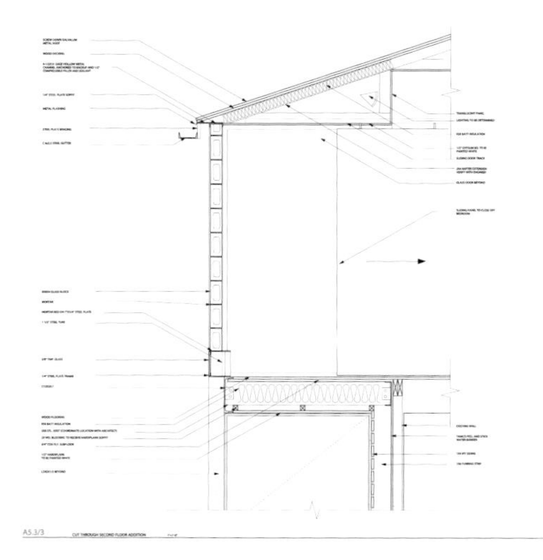

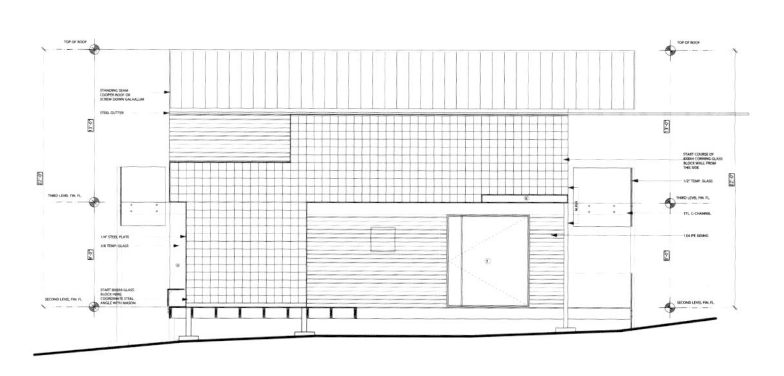

Project Name: Wissioming Residence
Design Company: FAIA Architecture
Design Company: Robert M. Gurney

项目名称：幽谷里
设计公司：FAIA 建工
设计师：罗伯特 M·格尼

层林尽染，四季入画

THE HOUSE DYED IN THE FOREST

美国马里兰州，丛林掩映，“回声”幽谷，波托马克河绕膝而过。而“回声”幽谷的声名在外，华盛顿特区郊区更有现代住宅以其命名。本案以原有基地为据，利用其固有低洼另辟小池，可谓大大减低对环境影响。

办公空间，位于一楼，旁依小池。独立建筑，居不出户，事业家庭两不误。其间附设车库，客卧套房。半透明的玻璃，品牌面板点燃激情。远远观看，灯火辉煌，如同主建筑旁悬挂的灯笼，令人遐思迩想。

结构性板块预先制订，被快速安装，加快了工期进程。结构设计打造液体循环加热及冷却体系，极尽环保手笔。釉面南墙，宽大挑檐。巨树华盖，阴阴夏木。

山水之间，滑动木质板面、镀铅锡钢、黑

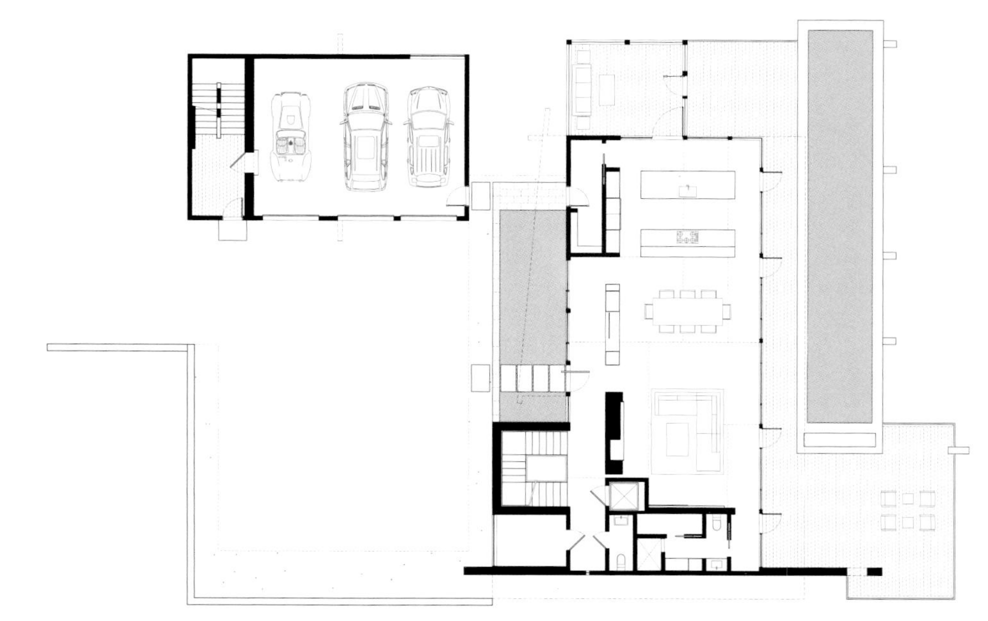

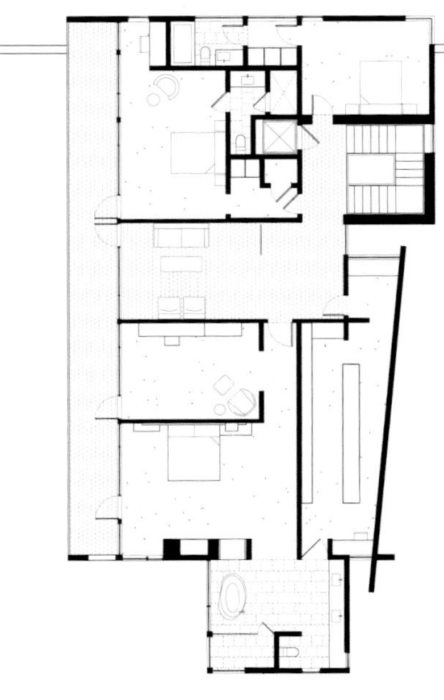

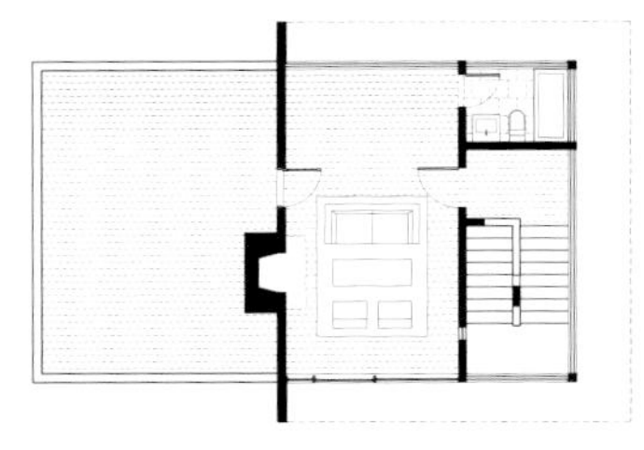

色钢窗，别样的户外调色板自然融入其中。另有青石、碎石点缀。
白色水磨石地板、白橡木橱柜、铝质板材提升光线华韵，精简细部设计。充足和微创详细的空间。或坐，或卧，或踱于如此静谧空间，欣赏于外面的美丽风景。设计的生命，得以绽放。

This house located in Glen Echo, Maryland is sited on a heavily wooded lot overlooking the Potomac River. Glen Echo stands as a rare enclave of modern houses in suburban Washington DC. The new house occupies the footprint of a pre-existing house in an effort to minimally disturb the site, removing no mature hardwoods in the process. A new swimming pool is suspended twenty feet above grade to further reduce the impact to the steeply sloping site.

THE PHAIDON ATLAS

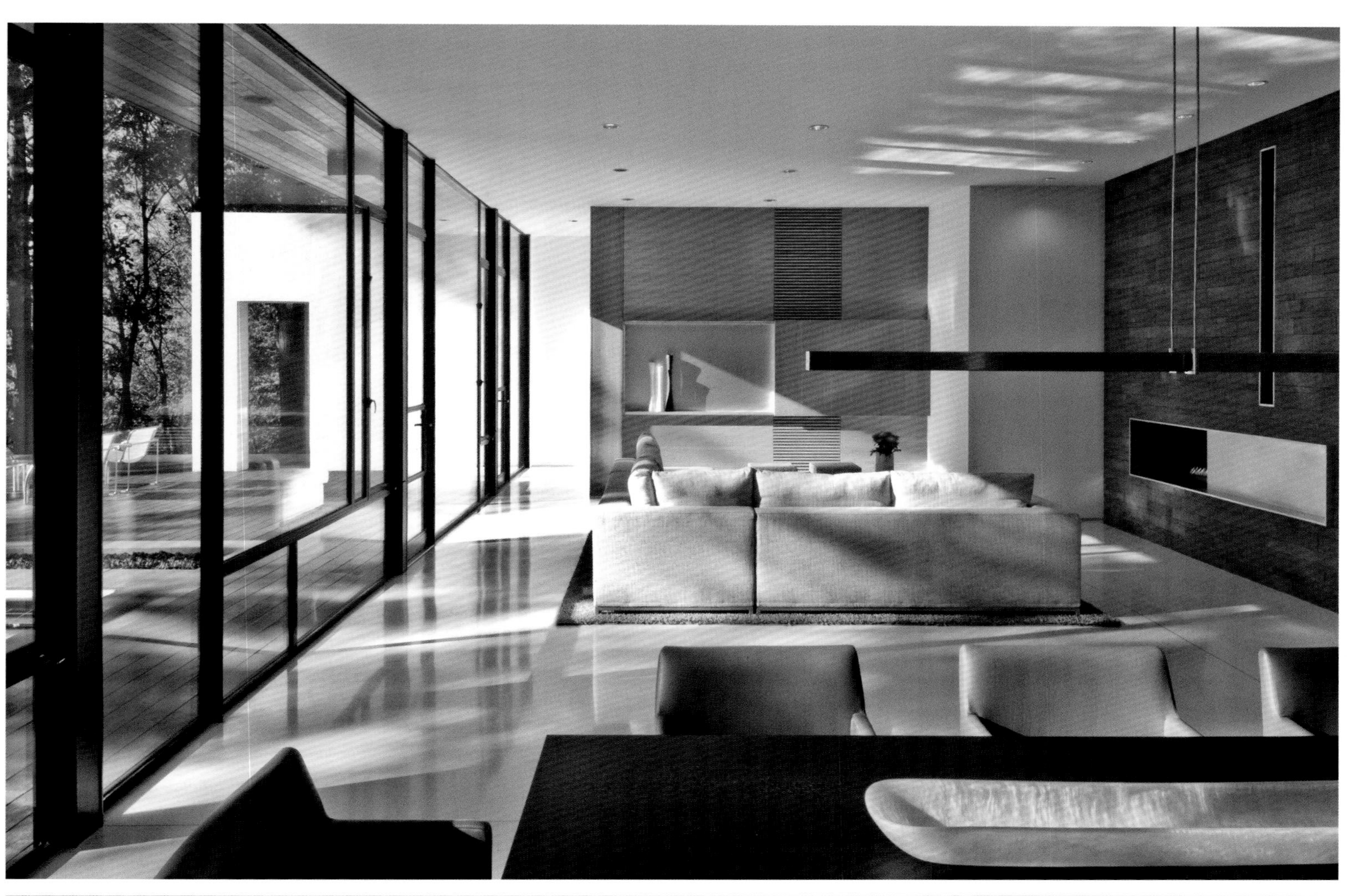

In an effort to lessen his dependence on the automobile, the owner builder required that his office become part of his residential compound. The office is located on the ground floor of a detached structure separated from the main residence by a reflecting pool. That structure also contains a garage on the first level and a guest suite above. Translucent glass and panels of Kalwall are used to allow the building to serve as a lantern to the main house at night.

Structural pre-cast concrete planks are employed throughout the project in effort to expedite the construction process, span large open areas and to provide the ability to heat the house hydronically. Combined with a 5" concrete slab and terrazzo flooring the structural system provides additional passive heating. Large overhangs on the glazed southern wall and the tree canopy minimize solar gain in the summer.

Wood siding is combined with soft gray terne coated stainless steel and black steel window frames to provide an exterior material palette that fits comfortably in the landscape. Bluestone, gravel and water complete the palette.

Interior materials such as white terrazzo flooring, white oak cabinetry and aluminum compliment the light filled and minimally detailed space. The creation of this atmosphere refocuses one' s attention outward, allowing the owner to reconnect with the inherently picturesque site already preserved through the design.

Project Name :House at Rua Alabarda (Alabarda Street)
Project Location: São Paulo, Brazil
Designer: Affonso Risi architect
Photographer: Paulo Risi

项目名称：House at Rua Alabarda (Alabarda Street)
项目地点：巴西圣保罗
设计师：Affonso Risi architect
摄影师：Paulo Risi

色彩点亮家之浪漫

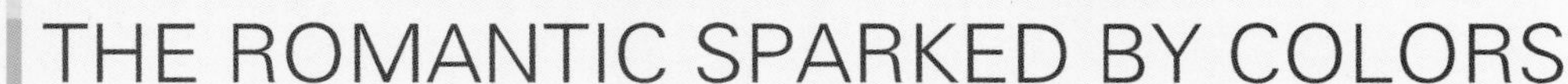
THE ROMANTIC SPARKED BY COLORS

这是一栋建在斜坡上的城市住宅，斜坡地面高于街道一米，斜坡上有一处通向车库以及工作区域的通道。斜坡层设有一个小水池，用于养殖鲤鱼。从斜坡向上利用一段加固的混凝土楼梯通往楼上，这里有一般住宅里所有的设备设施，如走廊、客房、厨房、阳台、办公室，另有三间卧室和三间浴室相连。设计之妙在于，在客厅和厨房之间还有一个日光浴室，由一段延长的楼梯连结接通了地面一块草坪和木甲板。另外该层的办公室和第三个卧室对斜坡底层尽头处的花园开放。

除此之外，为了空间设计的多样性，设计师利用空间原有的纵横方向，在鲤鱼池和楼梯之间设计了一个空隙，这样楼上的卧室空间和被粉刷成红色的楼下工作室就自然而然地构成了整个住宅的两部分。另外，设计师还在其他外部墙壁饰以白色，一方面为了使来客的视觉聚焦在房屋上，同时也使墙面和地面的衔接看上去更为和谐。

空间外部的其他铝制门窗也都漆成白色（静电环氧树脂），除了书柜是用钢筋水泥制成，其他室内地板也不例外。整个空间的设计遵循了黄金分割原理，使其更富有空间流动性和连续性。

Urban house in an upward sloping ground, narrow and long.

On the ground floor, one meter (3.23 ft.) higher than the street, there's an access, garage, studio and working areas. On this level , there's also a little pool filled with carps, from which a flight of stairs in reinforced concrete emerges, leading to the upper level, with all the house accommodations: access gallery, rooms, kitchen, verandas, office, 3 bedrooms and bathrooms. The office and the third bedroom open to the upper garden at the end of the ground.

There's a solarium over the living room and kitchen, with a lawn and a wooden deck, reached by the continuation of the stairs.

A void was designed in order to diminish the strong predominance of the longitudinal over the transversal direction, where the carp pool and stairs are located, separating the building in two bodies. The volume with bedrooms (upstairs) and studio (downstairs) was painted red. Other external walls are white. The purpose was to visually enhance the house and, at same time, to harmonically insert it on the ground.

External aluminum doors and windows are painted white (electrostatic epoxy).

Indoor floors have white cement plates and bookcases are made with ferro cement.

All measures are controlled by the Golden Section and the entire plan looks for fluidity and continuity of spaces.

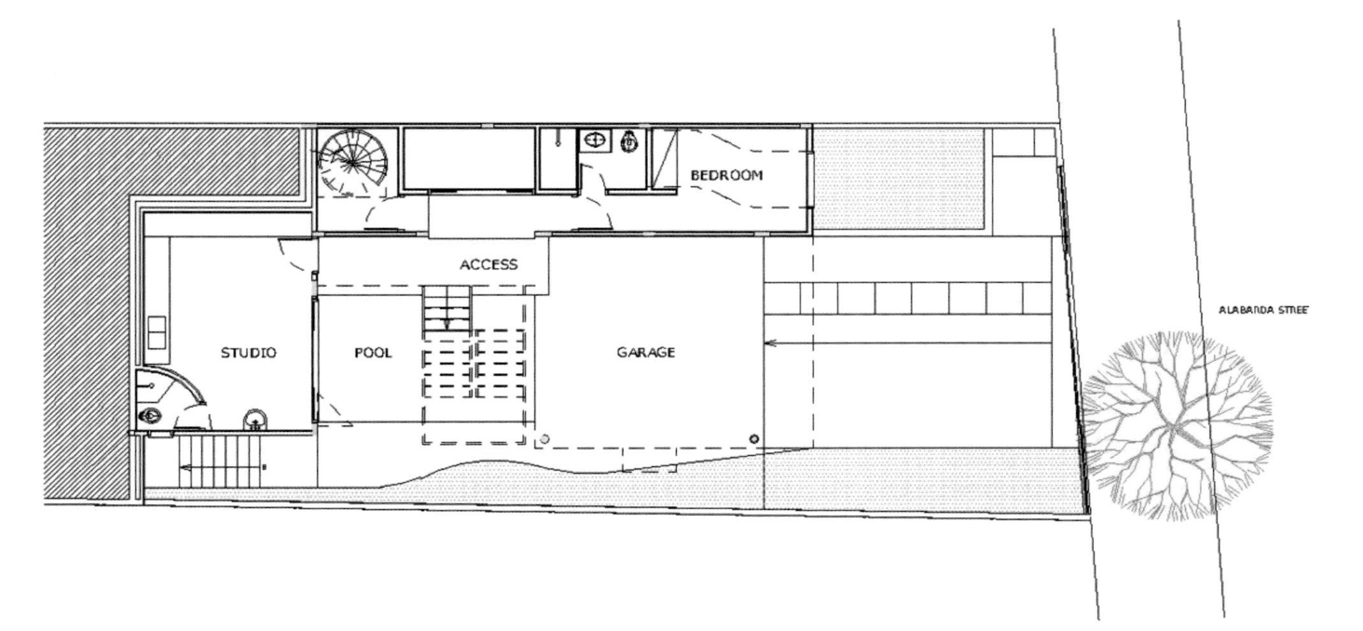
BEDROOM
ACCESS
STUDIO
POOL
GARAGE
ALABANDA STREE

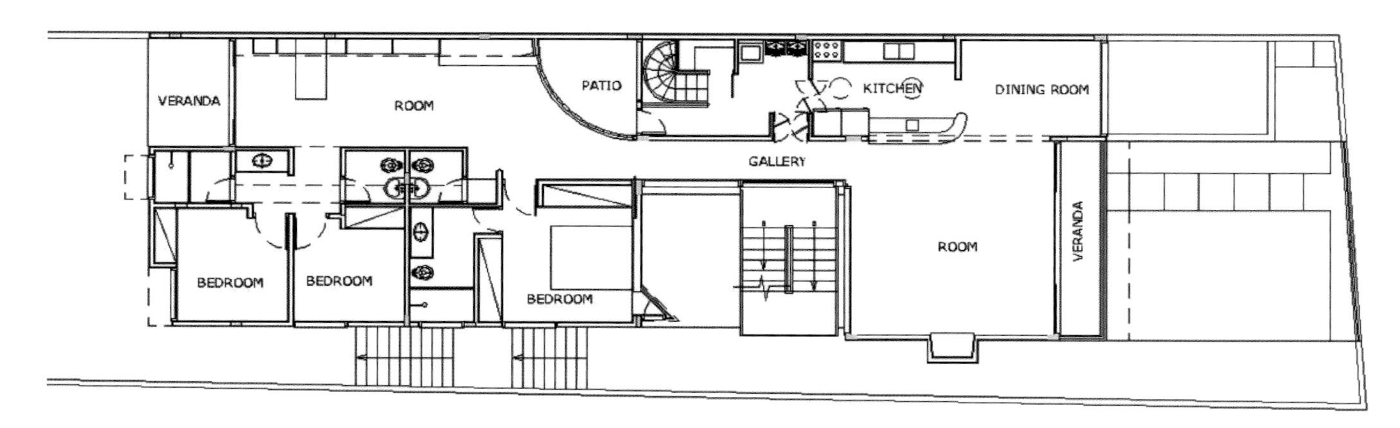
VERANDA
ROOM
PATIO
KITCHEN
DINING ROOM
GALLERY
BEDROOM
BEDROOM
BEDROOM
ROOM
VERANDA

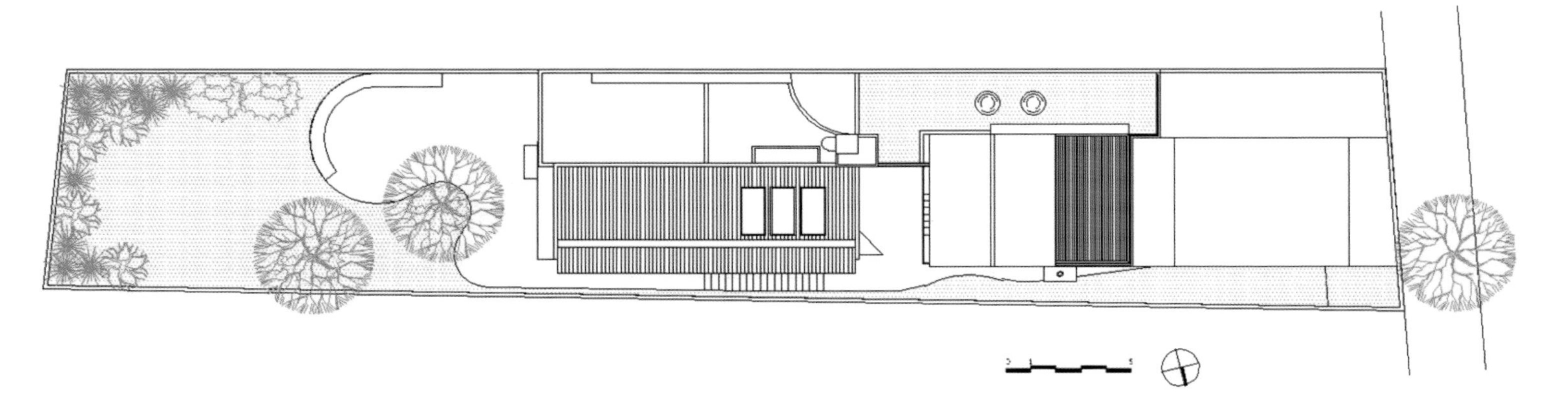

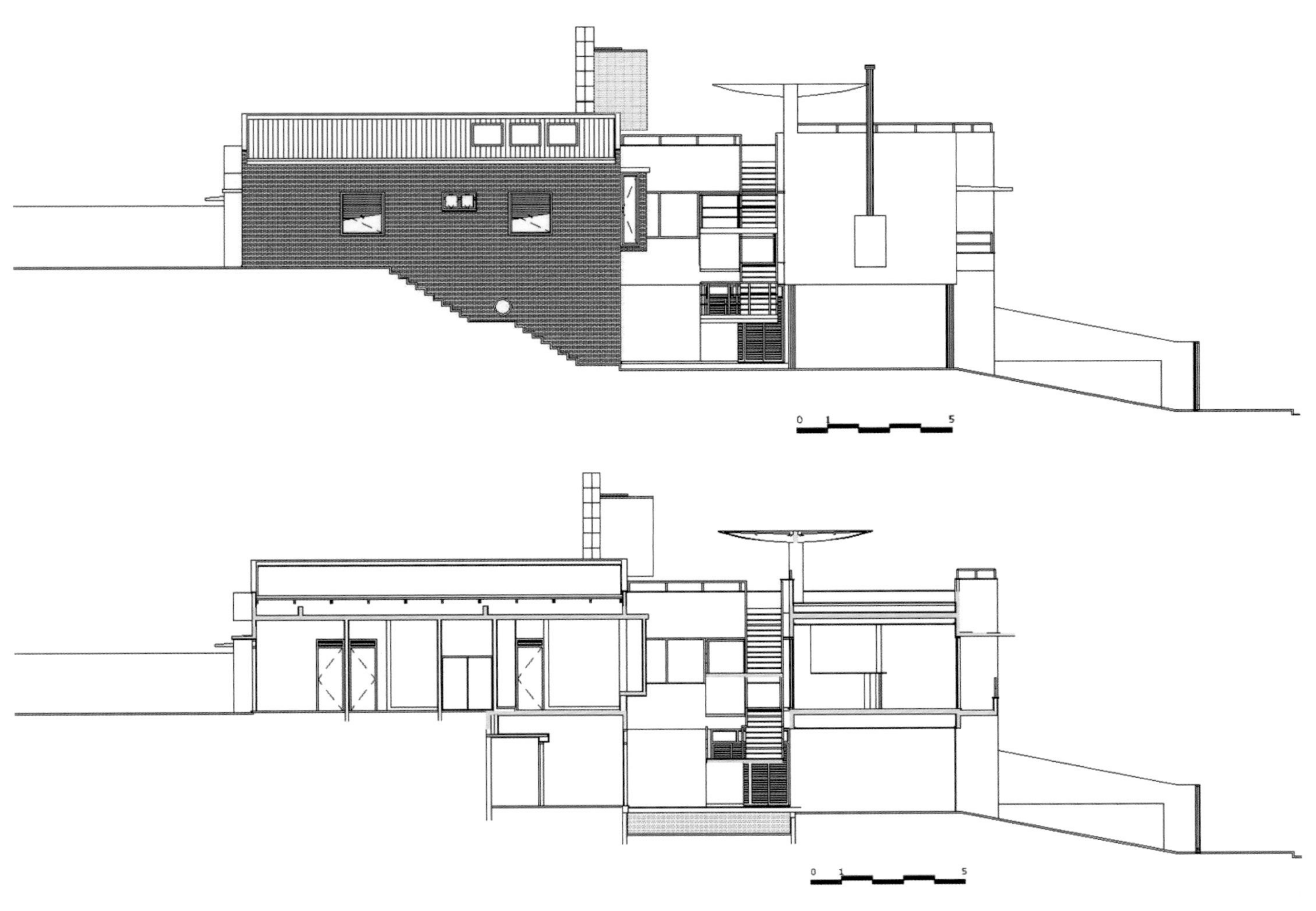
0 1 5
0 1 5

Project Name:Beach House
Location: São Paulo, Brazil
Arch. Affonso Risi
collaborator: arch. Willian T.S.Miyagui
builder: eng. Lineu Botta de Assis
photographs: Paulo Risi

项目名称：沙滩住宅
项目地点：巴西圣地亚哥
设计公司：Arch. Affonso Risi
合作公司：arch. Willian T.S.Miyagui
承建商：Lineu Botta de Assis
摄影师：Paulo Risi

面向大海，陶然雅居

THE SEASIDE HOUSE

该案建筑是几乎四面环海的五角规划楼，其中两层主楼共 258.8 m^2。二层共有五个区域角，空间设计允许楼上形成开放式设计，其中有四间房朝海，作为观海景的卧室；第五个区域角顶点作为一、二两层楼的过渡空间。另外，整个地面也设计成有滑动玻璃门的走廊。

生活区域如厨房、壁炉、浴室、衣橱、楼梯等，设计在每个区域角的中心，在其中使用加固混凝土结构构件和管道，以形成悬臂式五个顶点的房屋结构。空间所有的设计都遵循黄金分割和斐波那契级数法则进行。

The main building, a two floor pentagon plan pavilion (258.80 m^2/2 784 ft^2), is almost entirely surrounded by the sea.

The plan allows disposing 4 bedrooms upstairs opened as verandas to the sea landscape. The fifth vertex is an open space two stories high. The entire ground floor is also designed as a veranda with sliding glass doors.

Servant spaces (kitchen, fireplace, bathrooms, closets, stairs) are in the center of each face of the polygon and also contain the reinforced concrete structural elements and pipes, allowing the five cantilevered vertex of the house.

The entire design is controlled by the Golden Section and the Fibonacci Series.

There are two other pavilions as well: a square plan for employees and another one for guests near the swimming pool.

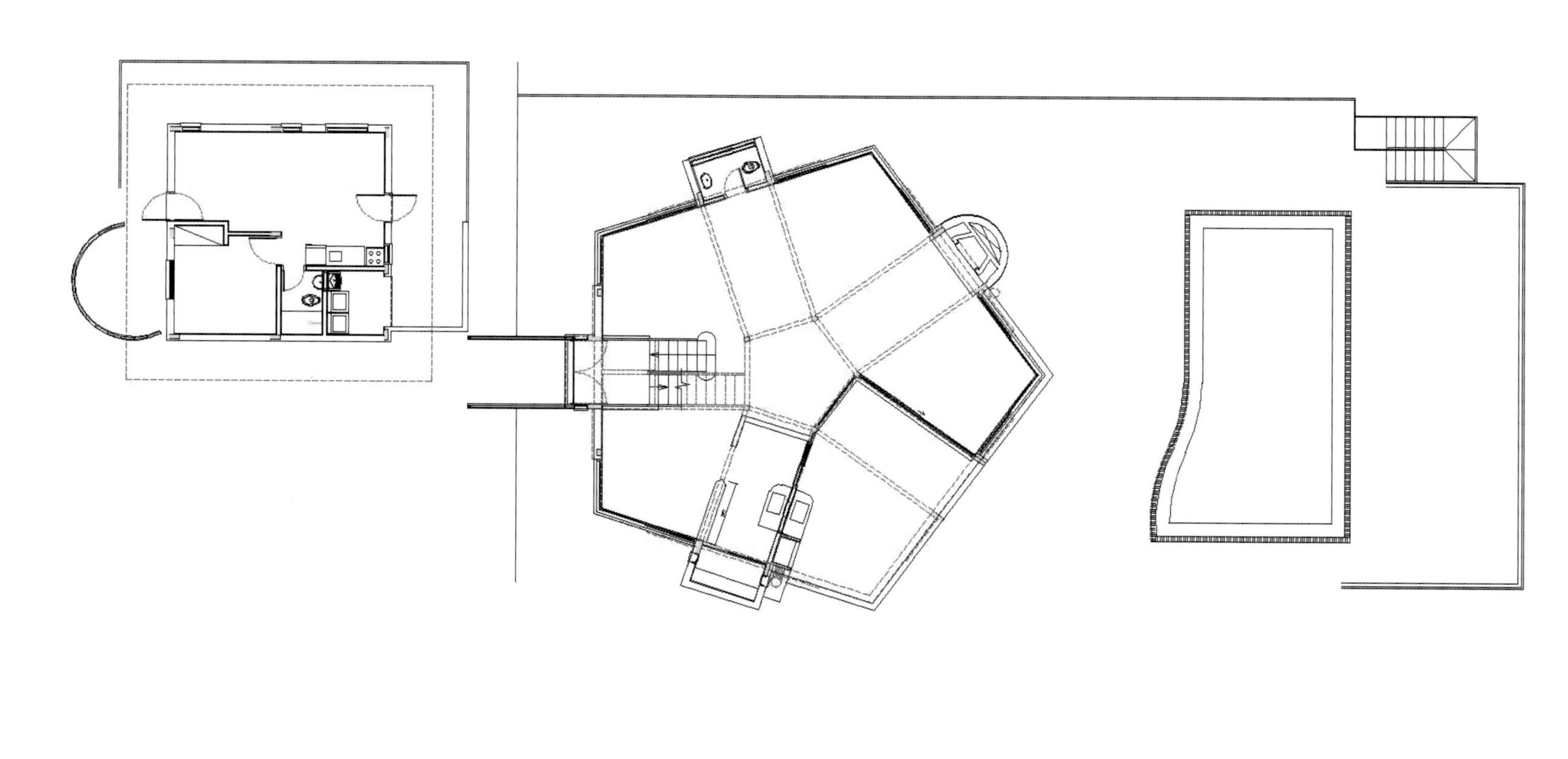

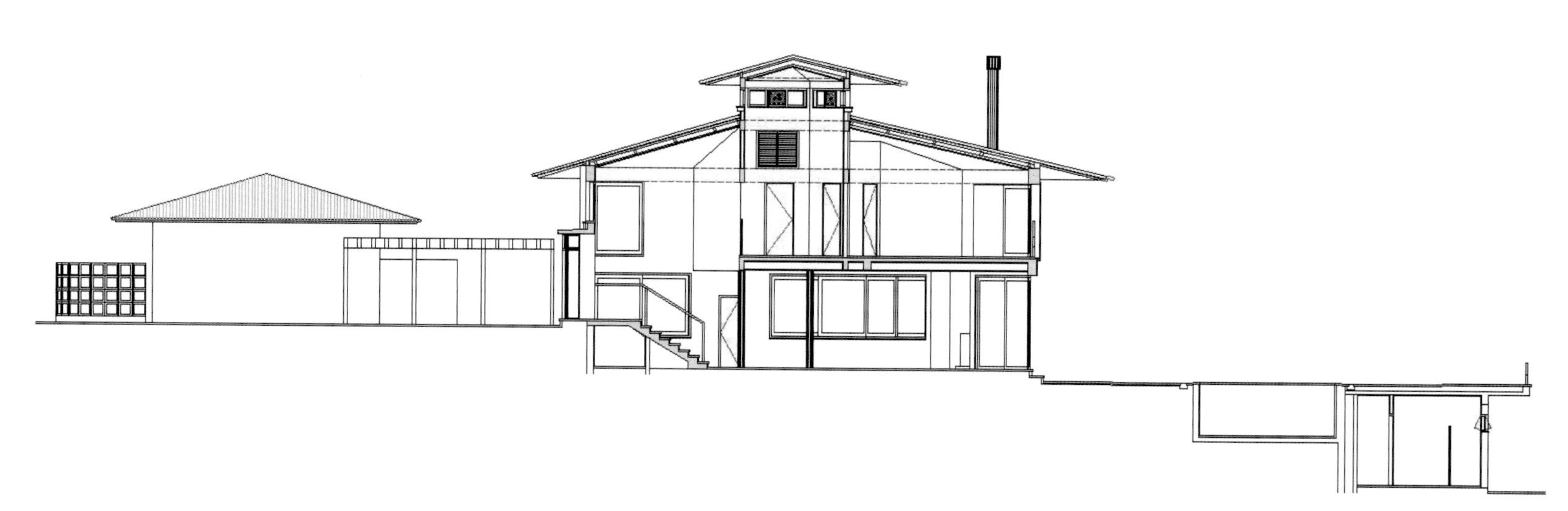

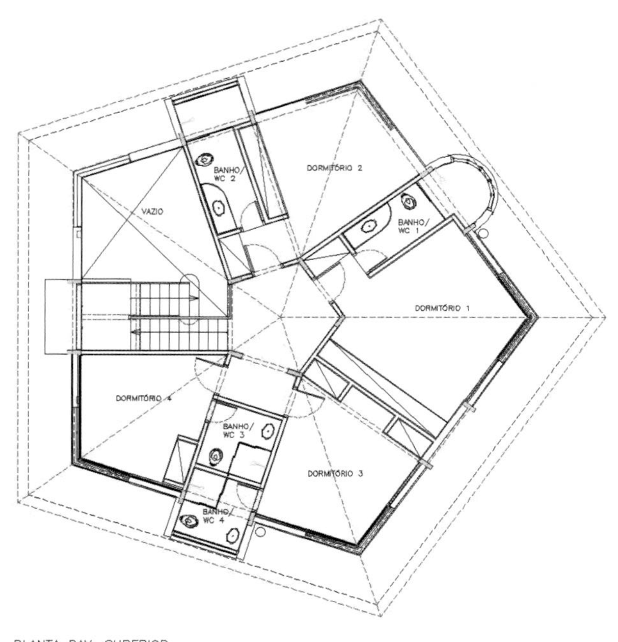

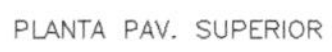
PLANTA PAV. SUPERIOR

Project name: Out of the Box
Location: Bangalore, India
Site area: 111.5 m^2
Design Firm: Cadence
Project team: Smaran Mallesh, Narendra Pirgal, Vikram Rajashekar
Photographer:Claire Arni
Materials: Cast in-situ cement blocks made from FRP moulds RCC sculpture in court finished with black oxide Black oxide flooring for stairs

项目名称：盒子之外
项目地点：印度班加罗尔
占地面积：111.5 m^2
设计公司：Cadence
设计团队：Smaran Mallesh, Narendra Pirgal, Vikram Rajashekar
摄影师：Claire Arni
主要用材：现浇水泥砖、黑色氧化物、铺板材料

光移影动，花样美宅
THE MANSION VARYING WITH LIGHT AND SHADOW

这间 120 m^2 的房子所呈现出的都市场景皆很经典，房子两旁紧连着普通民居，另外两侧则是保障房。面对此情此景，设计师不禁开始思考这间房子与外界环境所存在的紧密关系。

针对该案，设计师认为，最重要的是在阻隔屋外些许萧条景观的同时，也要做到融合公寓内外的环境。一般的做法是把房子设计成传统的庭院式住宅，而庭院的位置通常被房屋绕在中间。但设计师打破传统庭院式的住宅设计，大胆地把房子分为四个部分，并把庭院提升到二楼的西北角，营造出新颖空间的效果。围绕在庭院周边的起居室、餐厅和睡房不仅延伸了视觉，更与露天庭院相互呼应。设计师更是施以妙手，在庭院内添加了雕塑元素，长形的雕塑可作为餐桌，为户主提供非正式的用餐区域，而雕塑的凹槽里种植着绿树，如此一来庭院的使用功能被大大提高了。

为了建造一幅特别的 Jali 墙，设计师把庭院提升到房子的二楼。墙身的通孔由上往下减少，并在墙角交汇，形成了一个爱心形状。墙上的瓦片用 2x2 的纤维玻璃铸模浇铸而成，每块瓦片的厚度达 100 cm；铸模上有 4 个椭圆的通孔，设计师按照图形所需填充部分的通孔，造出想要的图案变化。加上庭院的墙身采用现浇 Jali 墙的设计，墙身的通孔增强了公寓的采光，有利于空气流通。

设计师设计的巧妙之处在于，不但为该案减低外界环境带来的负面影响，还不经意地把外面的景色带到了屋内。

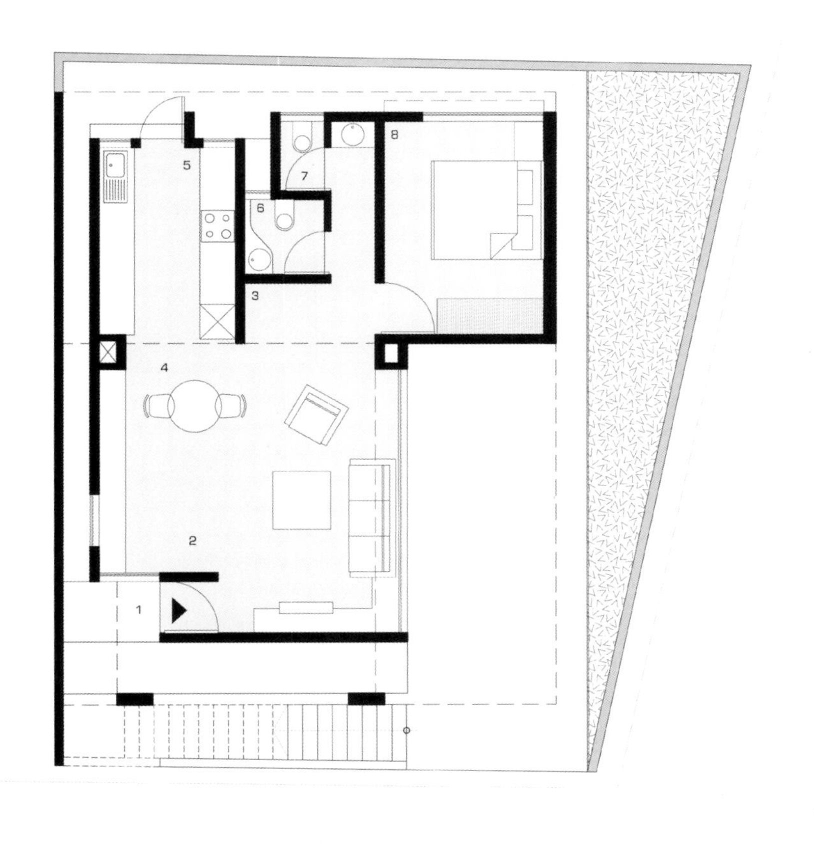
5
7
8
6
3
4
2
1

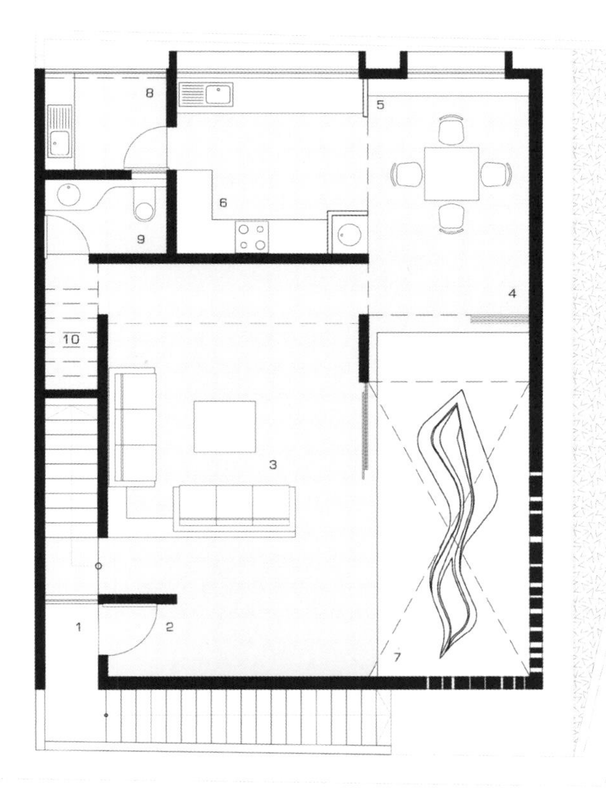
8
5
6
9
4
10
3
1
2
7

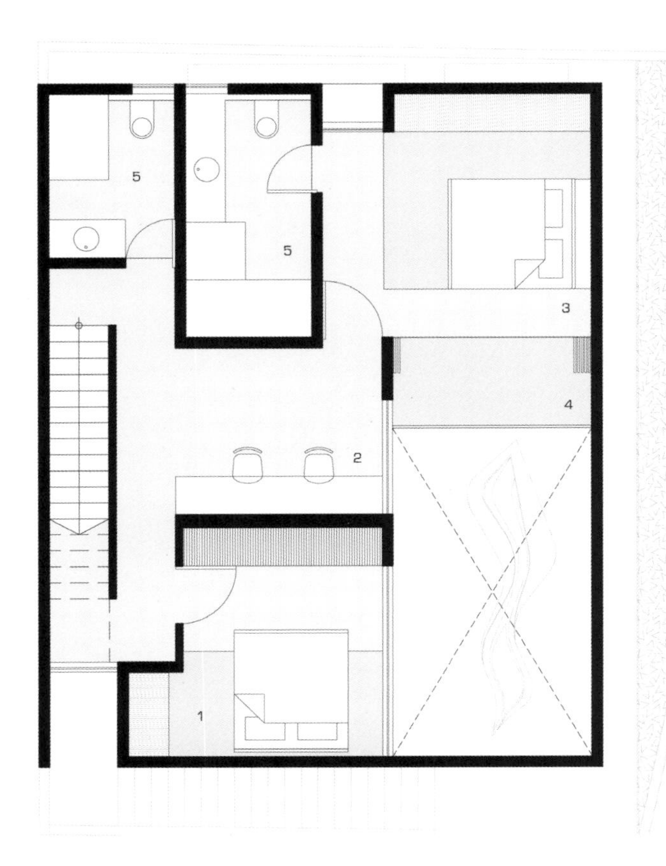
5
5
3
4
2
1

The 1 200-sq.-ft. corner site presented us with the classic urban scenario. The site was abutted by houses on two sides and flanked by low income housing on the other two sides. The question thus posed to us was, "What would be the relationship of the dwelling to the outside?" The stand taken by us was to incorporate the 'outside' inside while the building shuns the surroundings. A classic diagram of this would be the traditional courtyard house. Taking this classic diagram we moved the court to the corner to create new spatial and formal effects. By moving the court to the fourth quadrant of the square we could magnify the boundaries of each program flanking the court i.e. the living room, the dining and the bedrooms would not only feel much bigger but also would have sectional relationship with the open to sky court. The court is further articulated by placing a sculptural element that would serve as an informal dining area as well as a tub for housing a tree. The jali wall cast in-situ completes the fourth corner to accentuate the experience of the court.

This idea of the elevation was to have a customized jali of wall with a pattern of openings which fade away to form the platonic cube. A fiber glass mould of size 2' x2' was made to cast the concrete tiles; the thickness of this tile was 100mm. The mould had a set of 4 oval openings; these openings were then filled up according to the pattern to achieve the desired variation.

Thus the building while negating the outside environment simultaneously houses the 'outside' inside.

Project Name: Zen Modern Residence
Design Company: Forma Design

项目名称：禅的现代住宅
设计公司：福马设计

亚洲灵感，华丽转身
THE ASIAN-INSPIRED DESIGN

本案位于华盛顿特区，属亚当斯摩根地区。复式顶楼，殖民特点。亚洲灵感，华丽转身，空间独显宁静、禅韵之意境。

独立功能空间，如客厅、餐厅、厨房，在此汇集成一首和谐之曲。生活、餐饮、娱乐尽在其中。日式瓷石堆叠壁炉，两边各以定制橱柜收纳AV设备。一火烧赤红墙体横向穿越空间，恰成楼梯、壁炉之大背景。阳台空间，墙体延续火烧赤红色调，与内里空间合二为一。厨房内的先进厨柜、家电设施，功能实用，外观艺术。LED照明，应娱乐、烹饪、休闲肆意变化，独显生活的随心所欲和心情的自由自在。

楼上私密空间，两卧与画室、书房尽显生活、工作之便利。偶尔客人留宿，轻轻滑动定制书柜，客床跃然而出。主卧铁刀木床头，以LED壁龛为伴，一路引向主浴，内藏红外线桑拿。设计之精巧，思维之缜密，铸造空间无限神韵：诠释着禅而现代，现代而禅的空间。

A penthouse duplex in the Adams Morgan area of Washington DC was radically transformed from a dated Colonial, chopped up, developer grade apartment to a uber-chic Asian-inspired home for a well traveled busy professional who values downtime, serenity and meditation.

The separate living, dining and kitchen areas were combined into an open plan living/dining/entertainment stage. Japanese stacked porcelain stone envelopes the fireplace mass, and custom cabinetry on each side hide all the Av equipment. A burnt orange accent wall transverses the apartment and becomes the backdrop for the wenge stair mass and the fireplace, as it continues all the way to meet the burnt orange brick at the balcony outside. State of the art kitchen cabinetry and appliances facilitate the gourmet chef's every whim. From the LED lighting that changes with her mood to the natural materials throughout, the apartment is a playground for entertaining, cooking and lounging.

In the private areas upstairs, two small bedrooms were combined into her painting studio/study. When an overnight visitor arrives the custom bookcases slide open to reveal a hidden guest bed. The Master Bedroom's wenge headboard with the changing LED lit niche leads to the en-suite Master Bathroom, which features a hidden infra-red sauna. These well thought out yet subtle moves all add up to a zen-modern result for this repeat client.

Project Name: Ho Chao Sample House
Design Company: Mark Lintott Design
Designer: Mark Lintott
Size: 450 m²

项目名称：北京样板房
设计公司：Mark Lintott Design
设计师：林马克
面积：450 m²

花式鸡尾酒一般地绚烂多姿
COCKTAIL

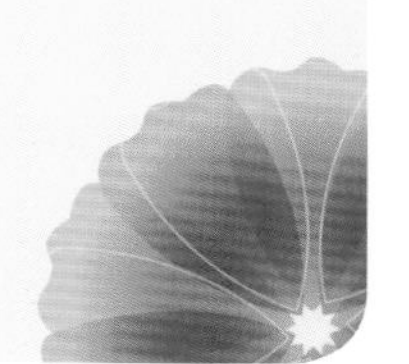

本案面积 450 m²，位居一楼。主接待室开阔、通畅，引领餐厅、厨房，并以贮酒墙界定玄关空间。另一端，窗户门楣通往花园露台及一年四季阳光普照的夏季房。前行即是较为隐蔽的主卧空间，如书房。另有三间卧室，一个 KTV 室、影音室及设备齐全的 SPA。

空间家具铺陈全部定制，彰显艺术品位，这些皆是本案设计用于其中国、欧洲市场之物品。

The apartment covers 450 m² on one floor. The large main reception room is open plan to the dining room / kitchen areas and screened from the main entrance by a substantial wine storage wall. Window doors at the opposite end of the reception room open onto a garden balcony and then to a "summer room" which has strong south light year round. This in turn connects to the private study which is part of the Master Bedroom suite. There are 3 further bedrooms as well as a KTV / AV room and a fully appointed spa bathroom en-suite to the master bedroom.

All furniture and artwork was custom designed and supplied to the project by MLD using resources both in China and Europe.

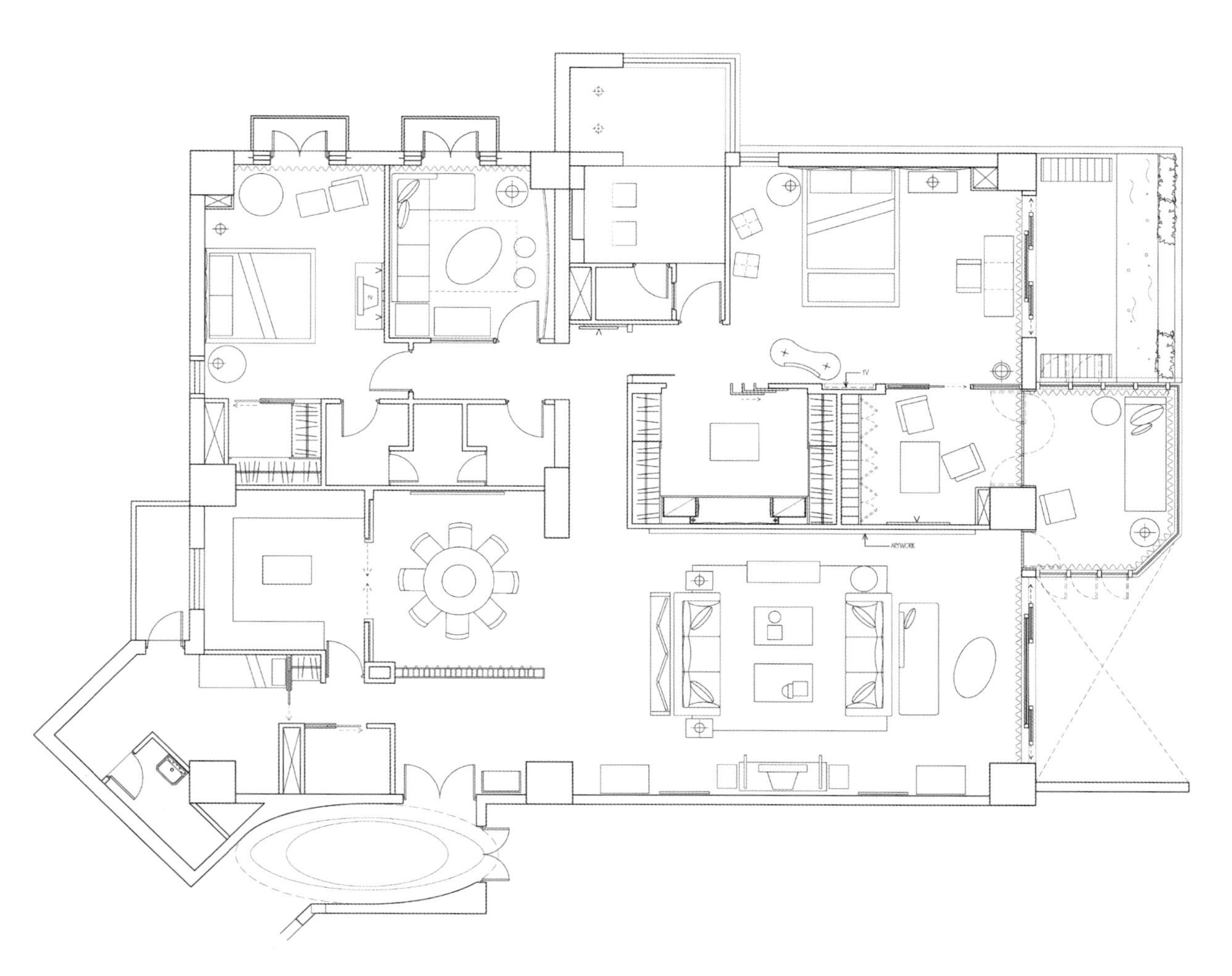

Project Name: NANKANG - TA TUNG Sample House
Design Company: Mark Lintott Design
Designer: MARK LINTOTT /FANG-WEI YU
Size: 148 m^2

项目名称：台北样板房二
设计公司：Mark Lintott Design
设计师：林马克
面积：148 m^2

红酒加咖啡，要让您沉醉
WINE AND COFFEE

本案位于台北南港，由美国建筑师史蒂芬•埃利希负责设计。
鉴于附近高科技中心人士为其目标客户群，设计自然以舒适、便利为题，尽显用材之新、科技之高。

设计中，原本计划中的两个卧室已为两个空间所代替。其一为女主人的更衣室；其二为男主人专用，一角专僻用作男主人专门的珠宝收藏。主接待室，大书桌后，原有壁龛，摆放经典的越野车、跑车车模。不同空间，不同设计，同样功能，相互呼应。

The sample apartment was part of a residential development in Nangang, Taipei designed by the American Architect Steven Ehrlich.

The development is aimed at buyers mainly from the nearby high tech centre in Nangang. As such we designed the apartment to have a classic comfort updated with the use of new materials and high tech installations.

The plan eliminates 2 of the proposed additional bedrooms giving the space over to a grand "boudoir" or dressing room/lounge for the wife while the husband has a collectors corner with jewel like models of classic sports and racing cars displayed in individual niches behind a large study desk in the main reception room.

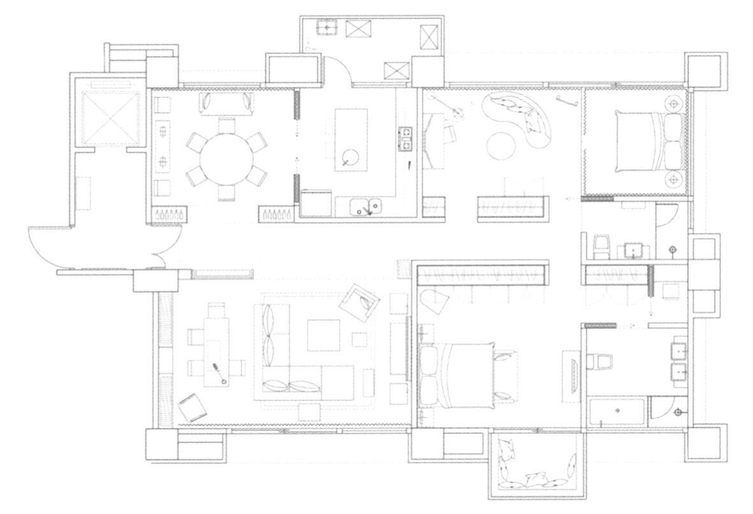

Project Name: SOTAI Sample House
Design Company: Mark Lintott Design
Designer: Mark Lintott, Chia Yen, Alison Yang
Materials: Mosaic, Wood, Acrylic, Wallpaper, Tile, Painting

项目名称：台湾样板房
设计公司：Mark Lintott Design
设计师：林马克等
用材：马赛克、木材、有机玻璃、墙纸、瓷砖、彩绘

摩登新人类的时尚新生活

THE NEW LIFE OF FASHION

本案样板房，应开发商要求设计，旨在为营销推广助澜、造势。约 200 m^2 的室内空间，原本设定为三间卧室、浴室，常规厨房，客厅和用餐空间。

重新设计后的空间减少卧室、浴室空间，扩大生活起居空间，并另辟娱乐空间。家居以年轻家庭为构想，生活起居空间便于旅游藏品及书本摆放，彰显旅游、娱乐、社交生活。

玄关独具特色，一墙用作美酒展示，一墙抛光木质镶板。地板，由石灰岩马赛克拼花形成，人字形图案。其他各墙、天花板延续空间彩绘主题。抛光木器、花卉图案，依然是传统的印花棉布壁纸所提供的灵感。

客浴用琉璃瓦镶嵌，雕刻浴缸、淋浴分区，界定明晰。主浴全白卡拉拉大理石。纯洁、华贵，呼应着其他空间的墙体、地板。

主卧毗邻客卧，附设书房，并专为孩子预留小小空间，该处也可作为另外一个学习空间。

The sample apartment is designed as part of the marketing programme for the select local property developer; Sotai Development. The interior covers an area of approx 200 m^2 and originally included 3 bedrooms and bathrooms with regular

kitchen, living and dining spaces.
MLD re-planned the apartment with one less bedroom/bathroom combination and put the released space into a larger living room and entertaining space.
The home is conceived around an imaginary family; young and international. Well travelled and social. The living spaces provide many opportunities for display of collected memorabilia and books. Places to work or to read and places above all for entertaining.
The entrance is framed on one side by a methyacrylate wine display wall and a polished wood panelled wall on the other. The floor is finished in a limestone herringbone pattern mosaic. All walls and ceilings are painted and the wood-work is polished and varnished with template floral patterns derived from traditional Chintz wallpaper patterns.
The guest bathroom is formed from a glazed mosaic with sculpted-in bath and shower areas. The master bathroom is finished entirely in white Bianco Carrara marble book-matched to all walls and floors.

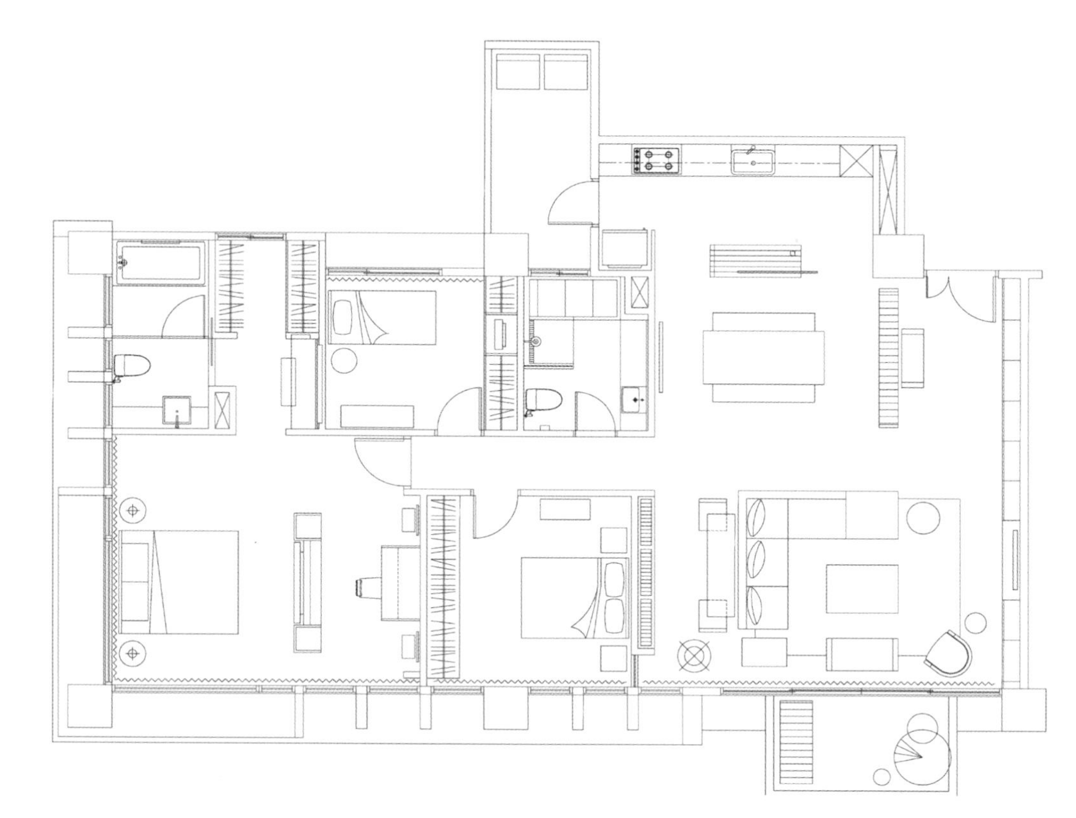

The master bedroom is combined with a study area in the room and adjoining this room is a guest suite and a further smaller room for a child or second study space.

SHAPING

形塑

别具一格，我形我酷，不平凡的艺术品味

UNIQUE AND DISTINGUISHED, ART-ORIENTED

Project Name:Bulthaup House
Designed By: K-Studio

项目名称：雪松居
设计公司：K 工作室

收纳家的奇思妙想
THE COLLECTION UNUSUAL AND WONDERFUL

moda bagno

1 000 m^2 的空间尽显设计手笔，由现代的经典家具铺陈，还有与众不同的卫浴、便利的厨房。
空间量体长 16 m，宽 14 m，高 12 m，典型的三维立体盒状结构。临街立面饰以钢板，穿插开口，真乃人间福地，家居生活中有蓝天、白云，白云蓝天下有怡然家居。雪松开口框架，温润阵阵，任其车水马龙，魅力自在空间，一窥其内之心不禁油然而生。
其一厨房，内有壁画。乃著名壁画艺术家乔安娜在希腊一鱼市临摹之作。

The 1 000 m^2 space hosts Moda Bagno and Interni's entire range of designer products. This includes contemporary classic furniture, bathrooms and kitchens.
The building is a large 12mHx14mWx16mD box. The street facade is clad with expanded metal and interspersed with openings to allow views into and out of the building. The cedar framework celebrates the openings and its intensified perspective form frames and gives direction to the views. As the cars drive by, the exaggerated perspective frames attract and intrigue. Their accentuated perspective allows for views from a wider range of angles, offering passersby more time to look inside whilst accelerating in their cars.
In one of the Bulthaup kitchens the mural artist Joanna Burtenshaw has drawn a scene from a Greek fish market.

Project Name: City apartments
Designer: Zhang Gang
Design Company: Zhang Gang Studio

项目名称：城市公寓
设计公司：张罡设计工作室
设计师：张罡

涌现自然
THE NATURAL SPROUTING

该案通过几何的组合和诗意的手段培育建筑系统之间连贯的关系，空间向着建筑中更高秩序整体发展，同时仍保持一种系统的表现，形成细腻野性的结构，结构的运用同时也重新诠释了家居空间。当结构从被压抑和再表现的苛刻状态中解放出来的时候，就能变得多样流动、色彩丰富，成为跨越类别、相互混合的一种结构生态空间。

本案中，设计师通过雕塑、浮雕一般的线条作为房间的主要装饰，通过线条的勾勒展现出野性与现代相结合的造型，让这套纯白的房子呈现出别具一格的简约风格，同时更具有后现代仿生特质。设计呈现出曲面的流线造型，蜿蜒伸展的全部墙面都是采用雕塑手法手工打磨而成，非常费时费力。其中，玄关的柱子是本案的设计亮点之一，原本是一根档路的承重柱，在设计师的巧妙构思下，变成藤树相绕的景观亮点。白色曲面在空间中交叠流转，又仿佛一朵盛放的花朵从中心柱体的花蕊向着周围的天花墙体恣意展开美丽的花瓣。

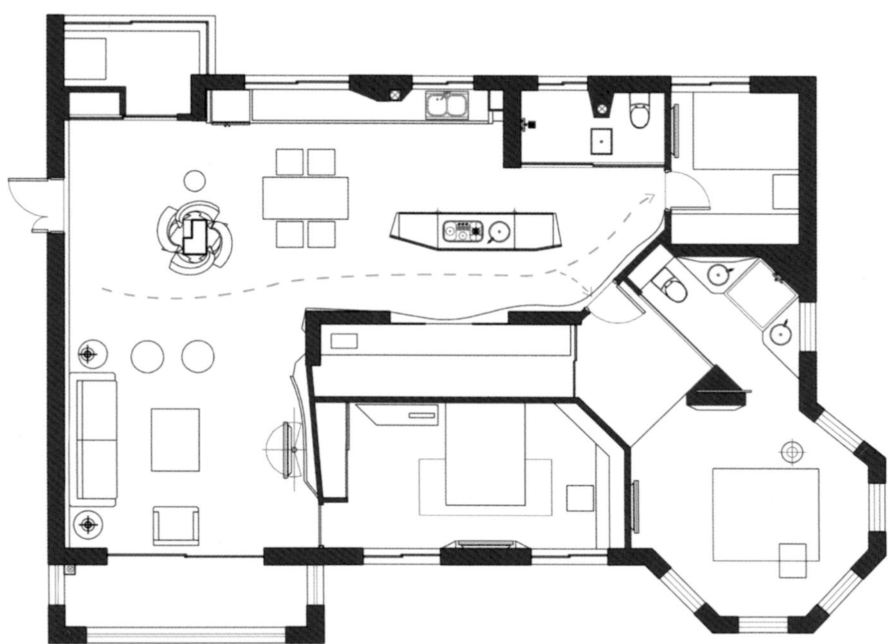

Combination of geometric and poetic means can nourish the coherence and harmony of building systems, a presentation of construction developing higher and more overall while still maintaining a fine original structure of systematic performance. The use of structure re-interprets the home space. When freed of the suppressed and represented, structure can be able to become mobile, variable, rich-colored, and to be an ecological space of mashup and of category crossing.

This is a space where elements of sculpture and relief are dominating and persistent, a space where lines brings out wild and modern styling, and a space of purely white showing a unique minimalist style, but affluent in more post modern bionic characters. The curves are certainly stream-lined; the stretches of walls are hand-polished treated in an approach of sculptural techniques; the columns in the entrance are one of the highlights when each looks like for creeper or vine climbing upward. Overlapping and interweaving in the air are white surfaces, like flowers in bloom high in the central cylinders to confide in their confidence to the wall and the ceiling.

Project Name: City apartments
Design Company: Zhang Gang Studio
Designer: Zhang Gang

项目名称：城市公寓
设计公司：张罡设计工作室
设计师：张罡

无界
NO LIMITATION

如果从服装设计跨界到家具设计，是让使用者体验一种更自然与舒适的生活氛围，那么从建筑设计跨界到室内设计，其重点则是给予了居住者一种更为宏伟、大气和有空间感的生活方式。在本案中，空间的形成，源于对未来的思考；功能的模糊，打破了对传统设计的思考，跨界在空间的各个要素中，达到一种全新的感觉与体验。

建筑与室内、室内空间与家具、居住与办公、现代与未来、对称与非对称、平面与立体、水平与垂直，关于空间与设计等二元概念，在这里，设计师希望超越传统与现代之间的对立、各种思维能够碰撞、产生火花并进行融合。于是，所有这些元素结合成一个奇妙的、激动人心的有机体。

连通一、二层的楼梯空间是一个中心节点，白与黑的虚实交错，奏响了线性乐章。楼梯踏脚的白板翻折回转，界定出的不只是一个楼梯空间，一边的低处踏板向着餐厅延展开去，另一边的高处踏板干脆转角横走，便构建出一方长条书桌。书桌尽头的墙体上，黑与白以曲线相隔，是墙身书柜，更是中国的水墨山水意向。

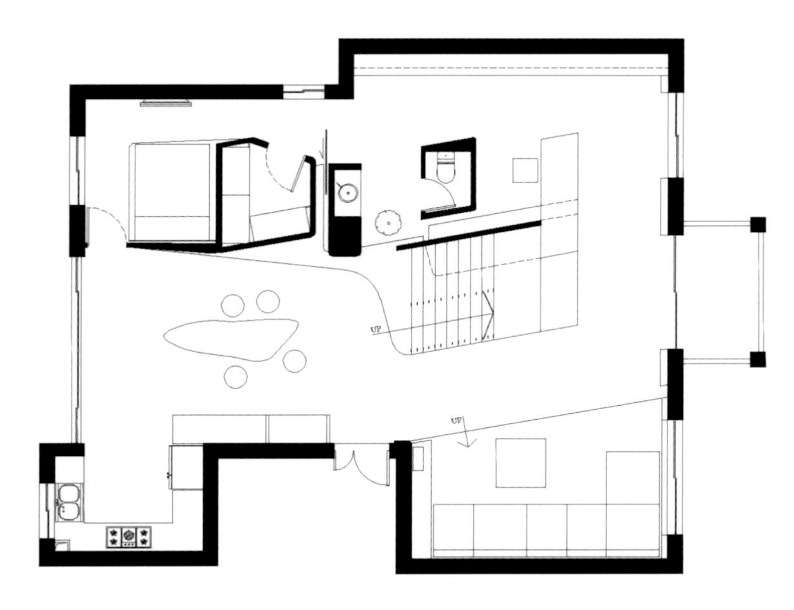

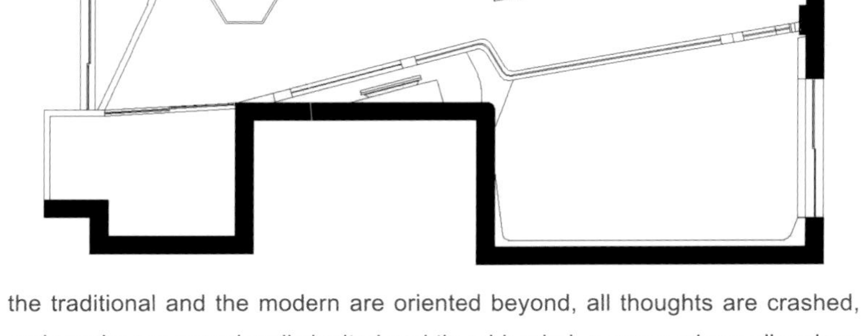

If design from fashion to furniture is intended for an atmosphere more natural and comfortable, then design into the interior across architect is bound to allow for a lifestyle, more magnificent, grand and filled with a spatial sense. This space as for the volumetric development inspired by thought and perception into future and blurred function, and breaking away from the traditional sights into design, takes access to varying elements for a completely-renewed experience.

Everywhere are contrasting concepts, like architect and interior, interior and furniture, dwelling and office, modern and future, etc. A natural result it is, where the traditional and the modern are oriented beyond, all thoughts are crashed, and sparks are occasionally ignited and then blended, a space where all make a marvel and exciting body.

The landing from the 1st floor to the 2nd is a central knot, where the lower footsteps lead to the dining room, while higher ones winding along the edge, accomplishing a long table. At the end of the table, are blocks of black and white separated by curves, not only bookcase on the wall, but also images of Chinese landscape paintings.

Project Name: City apartments
Design Company: Zhang Gang Studio
Designer: Zhang Gang

项目名称：城市公寓
设计公司：张罡设计工作室
设计师：张罡

墨域
DARK LAND

这是一间只有 60 m² 的个人公寓样板房，适合自由创意的 SOHO 一族或是作小型的工作室。设计师的思考由此出发。在富有思想和空间创造力的工作方式下，传统的空间概念已经不存在了，空间自然景观和现代生活成了设计的元素。人们更期待自然赋予的设计。

在本案的空间里，设计师运用了一些“随形”和流动的设计，来流露出人们更期待贴近自然的品位，试着提醒人们如何越过山丘、洞穴、河流，唤醒了人们追求自然天成的归属感。墙体上的“山形”造型书架，像磁石一样紧紧吸住人的眼球，这完全是现代的手法，但是却发出再自然不过的响声。样板房内唯一的绿色系物品——沙发凳上的垫子，犹如弯曲的山谷形状，与山形书架形成完美的呼应。在纯粹的黑与白两色协奏的公寓里，这点睛的一抹春绿竟带来分外的温柔。

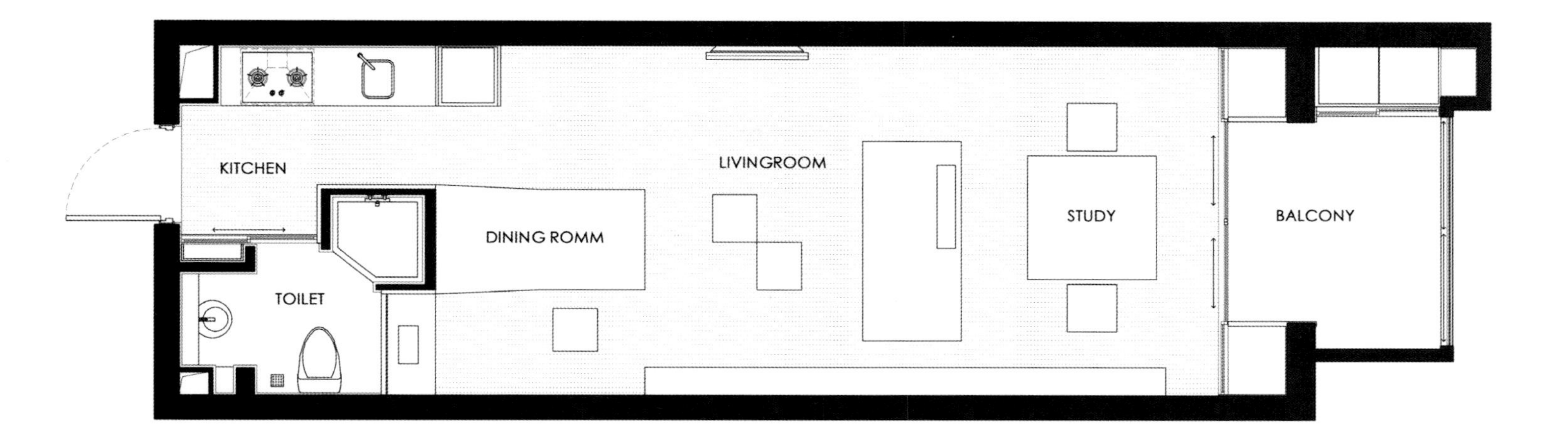

Despite its narrow space, this space is really an ideal place for SOHO or used as a small studio. Proceeding from such an idea, Mr. Zhang successfully breaks away the traditional spatial concept to design a space where landscape and modern life serve as dominating elements, a space well catering for expectation of a projected endowed by nature.

In the space, designs shape-following or fluid, reveals that people have a preference for taste getting close to nature, reminiscent of how people climb over hills, and go through caves or across rivers, and waking up belonging sense that human beings are in the pursuit of accomplishment by nature. The mountain-shaped bookshelf on the wall is like a magnet tightly drawing people's eyes. The sofa cushions, the only items in green hue, are designed into a valley appearance, and making a perfect echo with the bookshelf. It' s the very green that brings a sense exceptionally gentle and warm to the space in pure hues of black and white.

Project Name: Collector' s Loft
Design Company: UN Studio
Designer: Ben Van Berkel, Marianthi Tatari, Collette Parras
Size: 550 m^2

项目名称：藏宝楼
设计公司：UN 工作室
设计师：波克利、塔塔里、帕拉斯
面积：550 m^2

行云流水般的艺术之旅
THE FLOWING JOURNEY OF ART

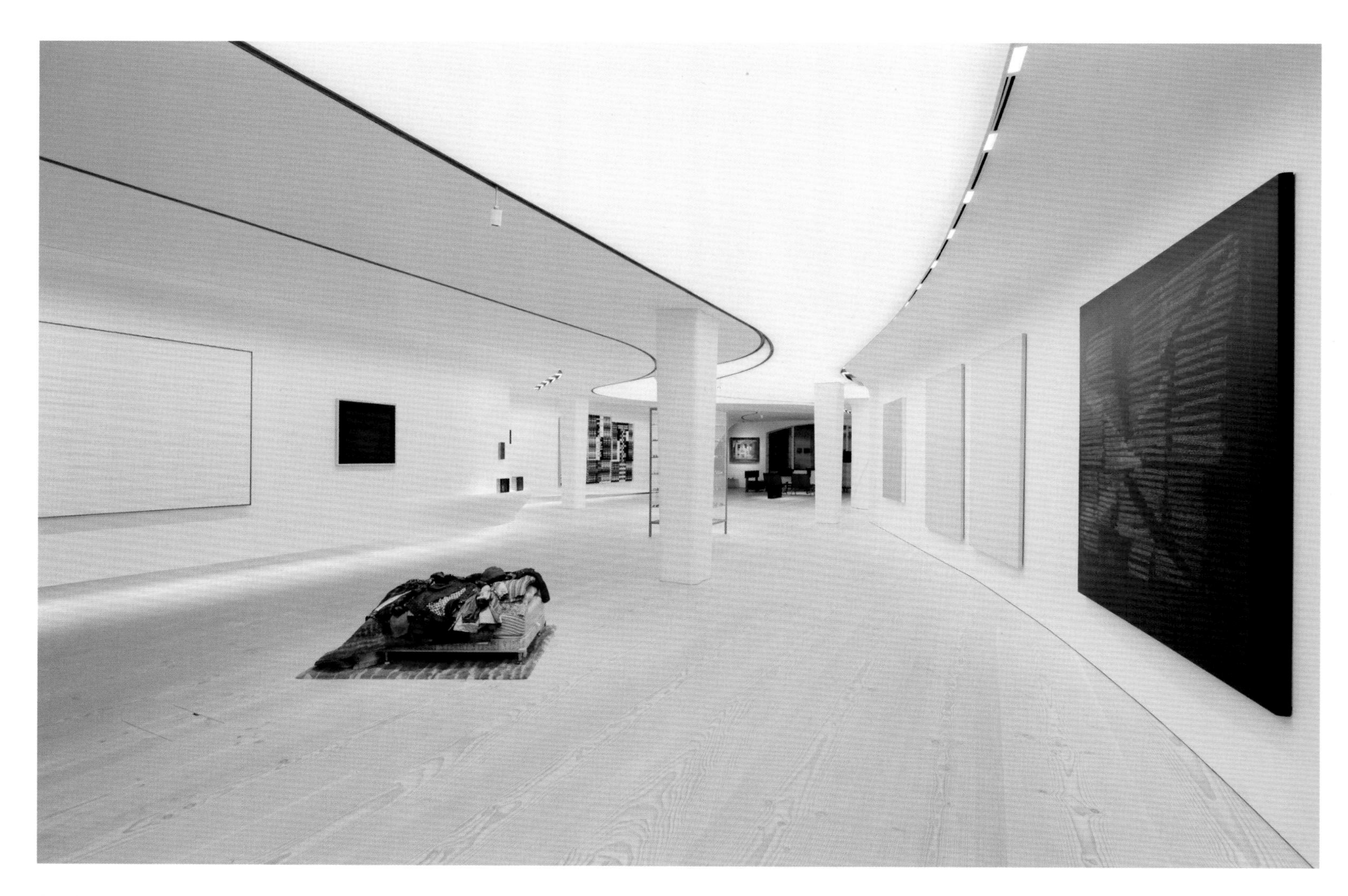

曼哈顿格林威治村，生活空展，画廊质感。游动主墙，铰链式天花，相互呼应。艺术展品，尽在生活起居、生活点滴。艺术典雅，墙面展现。天花或明或暗，灯光相宜，洒下一地华彩。南墙原有开窗依然不再，落地大窗，格子玻璃，曼哈顿城市风光，自然入内。

The design for an existing loft located in Greenwich Village in Manhattan explores the interaction between a gallery and living space. The main walls in the loft flow through the space, and together with articulated ceilings create hybrid conditions in which exhibition areas merge into living areas. While the walls form a calm and controlled backdrop for the works of art, the ceiling is more articulated in its expression of this transition. By interchanging luminous and opaque, the ceiling creates a field of ambient and local lighting conditions, forming an organizational element in the exhibition and the living areas. In addition the former windows in the South wall have been replaced by floor to ceiling glass panes that frame and extend compelling views, over a full glass balcony, toward downtown Manhattan.

18.JUN.1989
19.JUN.1989
20.JUN.1989

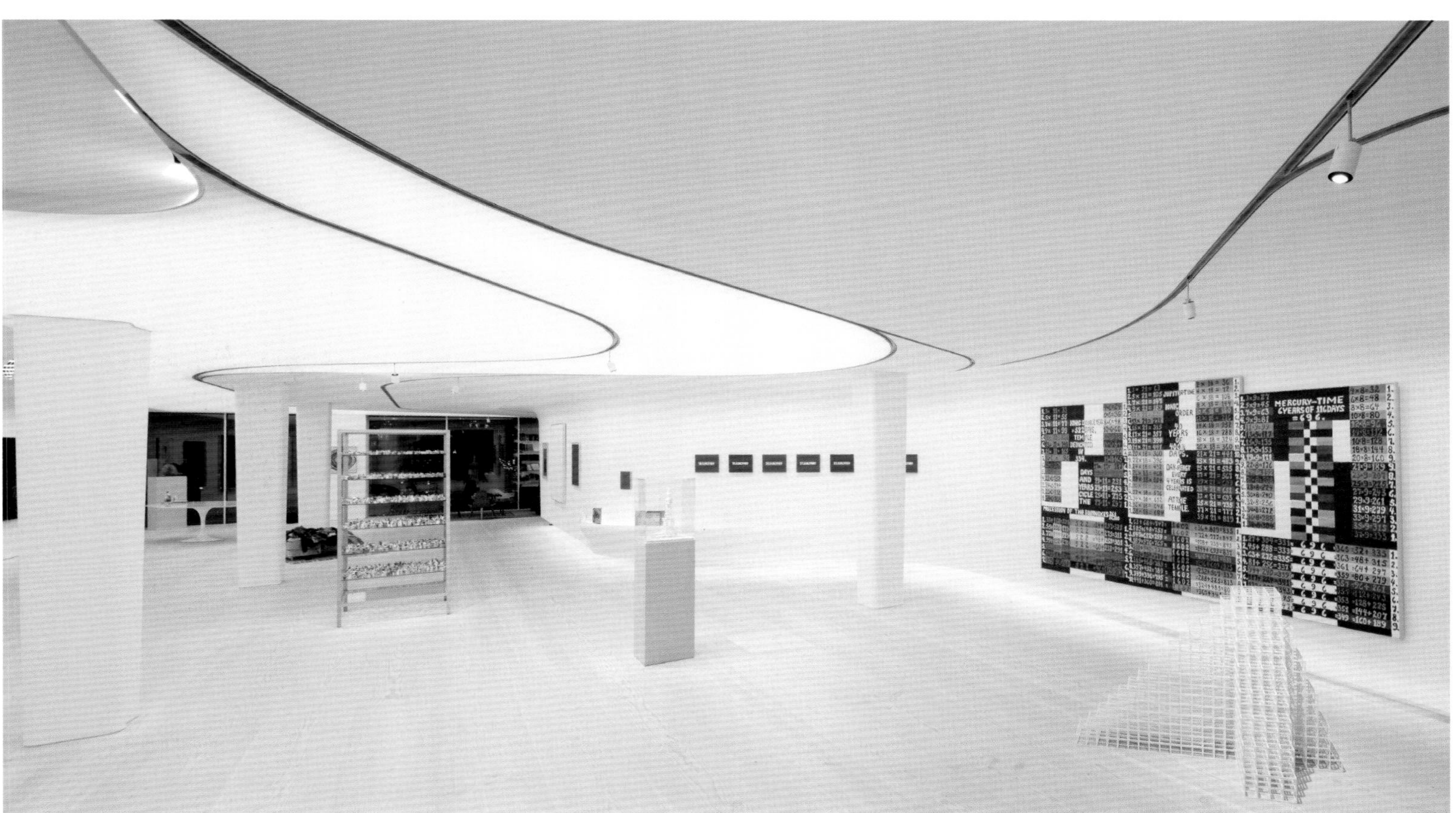
MERCURY—TIME

BERTRAND RUSSELL
BERTRAND RUSSELL

OCT.23,1989

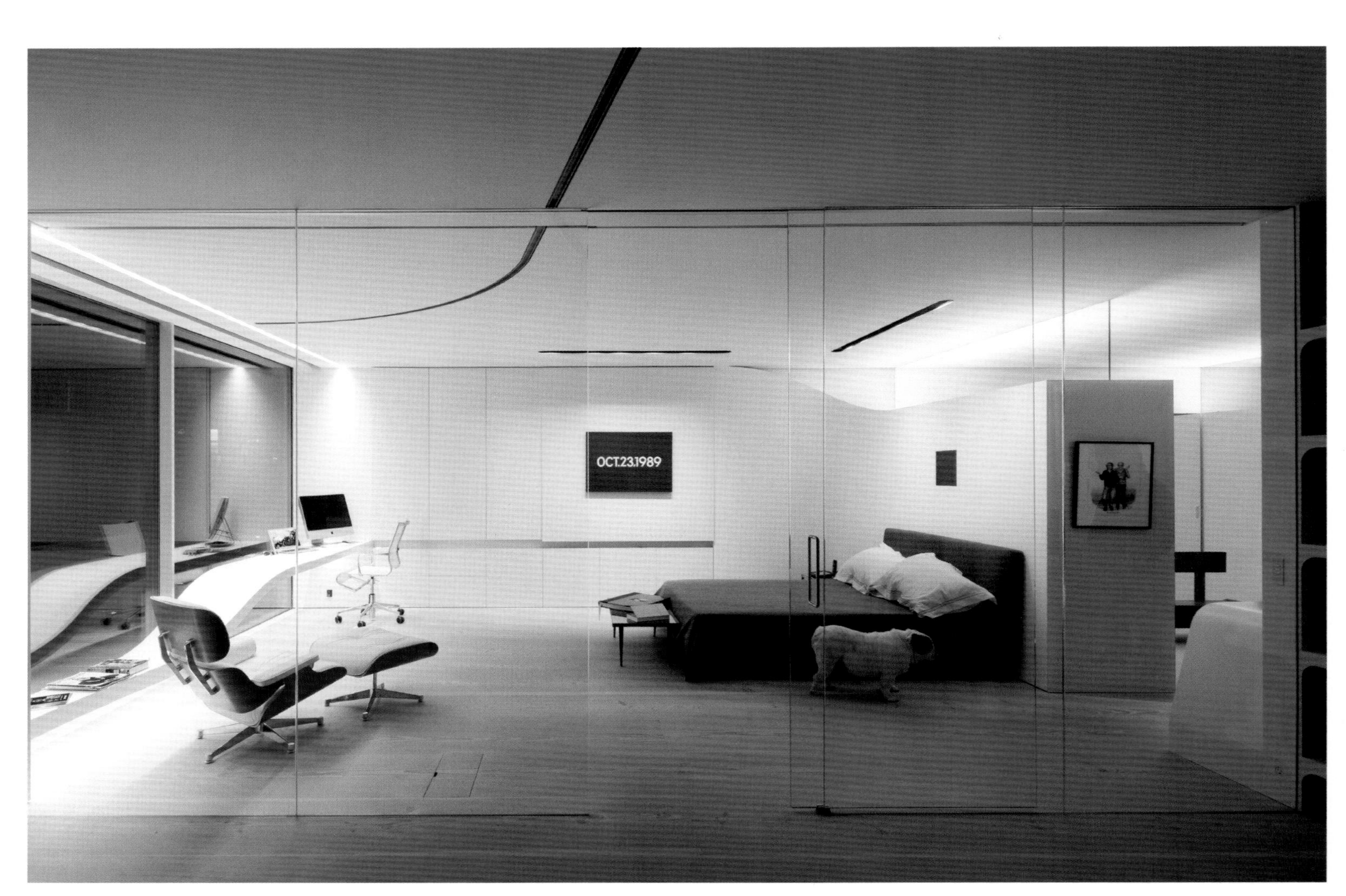
OCT.23,1989

23,1989

Project Name: Molding Show House
Design Firm:One Plus Partnership Limited
Designers Involved:Mr. Ajax Law Ling Kit & Ms. Virginia Lung
Type:Show House
Area:360 m^2
Materials:Carpet, curtain, sheer, leather, fabric, mirror, glass, laminated glass, laminated mirror, marble, corian stone, mosaic, plastic laminate, paint, metal in powder coating finished, stainless steel, aluminum, tiles, mother of pearl, timber floor, veneer, wallpaper
Photographer: 罗灵杰 Ajax Law Ling Kit

项目名称：武汉别墅样板房（B1 户型）设计说明
设计公司：壹正企划有限公司
设计师：罗灵杰、龙慧祺
类别：别墅样板房
面积 :360 m^2
主要用材：地毯、窗帘、窗纱、皮料、布料、镜、玻璃、夹纱玻璃、夹纱镜、云石、仿石、云石马赛克、防火胶板、油漆、金属喷粉、不锈钢、铝金属、磁砖、贝母、木地板、木皮、墙纸
摄影：罗灵杰 Ajax Law Ling Kit

立体花线塑造酷形酷态

THE SOLID

这所别墅样板房以“立体花线”为设计主轴，对称的花线设计是欧陆式设计不可或缺的元素，但在天花、墙身以及地脚位置加上对称花线的设计又过于传统，因此设计师为“花线”注入全新的元素，使之变得立体化，更让立体“花线”散布于整间别墅的不同位置中，这种创新大胆的设计意念，不单提升了视觉效果，更让整幢别墅样板房充满生命力，华丽高尚的格调打破传统浮夸的欧陆装潢，空间处处充满时尚感。

整幢别墅以黑、白、红为主要颜色，客厅及饭厅以白色为主，除了用上白色云石外，于天花及墙身设计了不同长度的立体花线，设计师没有就花线的定位作刻意对称的安排，这样反而更显空间感；加上巧妙地利用花线的线条组合而成灯壳，设计既简约又时尚。楼梯把手的设计配上造型特别的椅子等家具，不但贯彻别墅样板房西式的设计风格，更带出高尚的王者气派。

另外在饭厅中的屏幕，因为用上白色透视屏幕，以重叠花线砌成的图案，并没有一般墙体的压迫感。主人睡房的特色墙是另一设计亮点，此特色墙身是利用大量的切割木板拼凑而成，堆放一片片的立体花线木板而成为一完整的花线壁板，当中更设

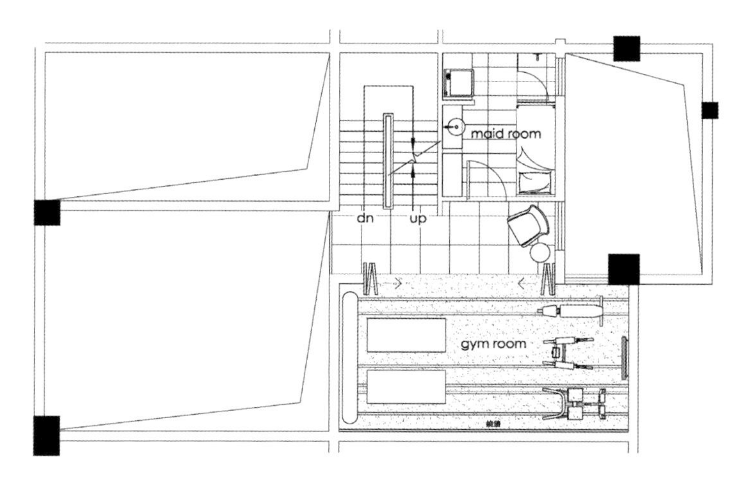
maid room
dn
up
gym room

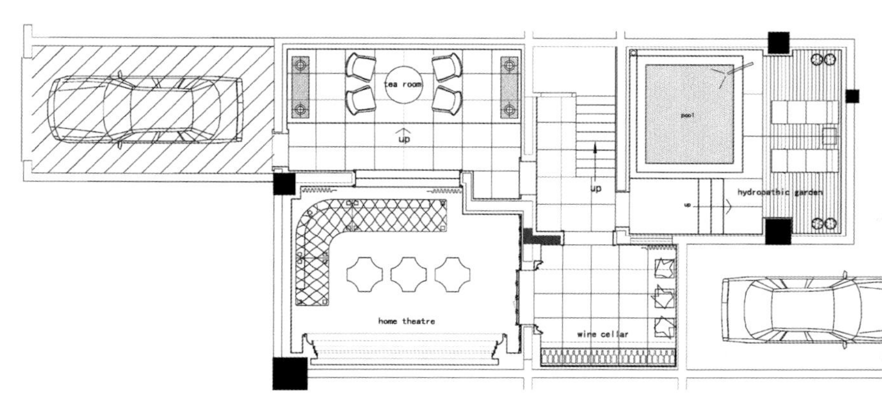
tea room
up
up
pool
up
hydropathic garden
home theatre
wine cellar

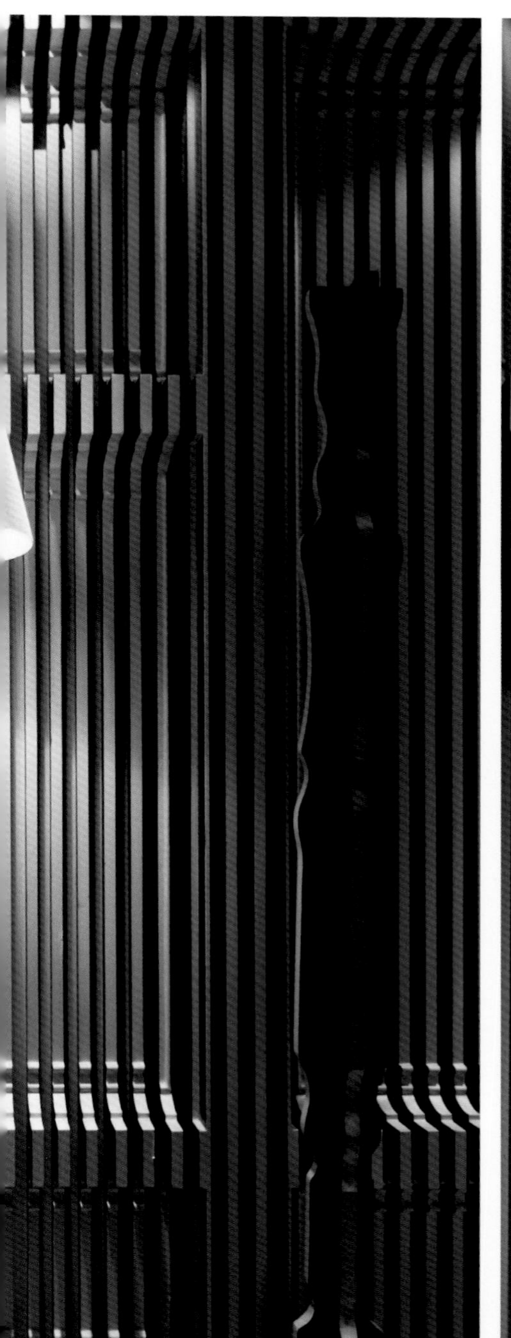

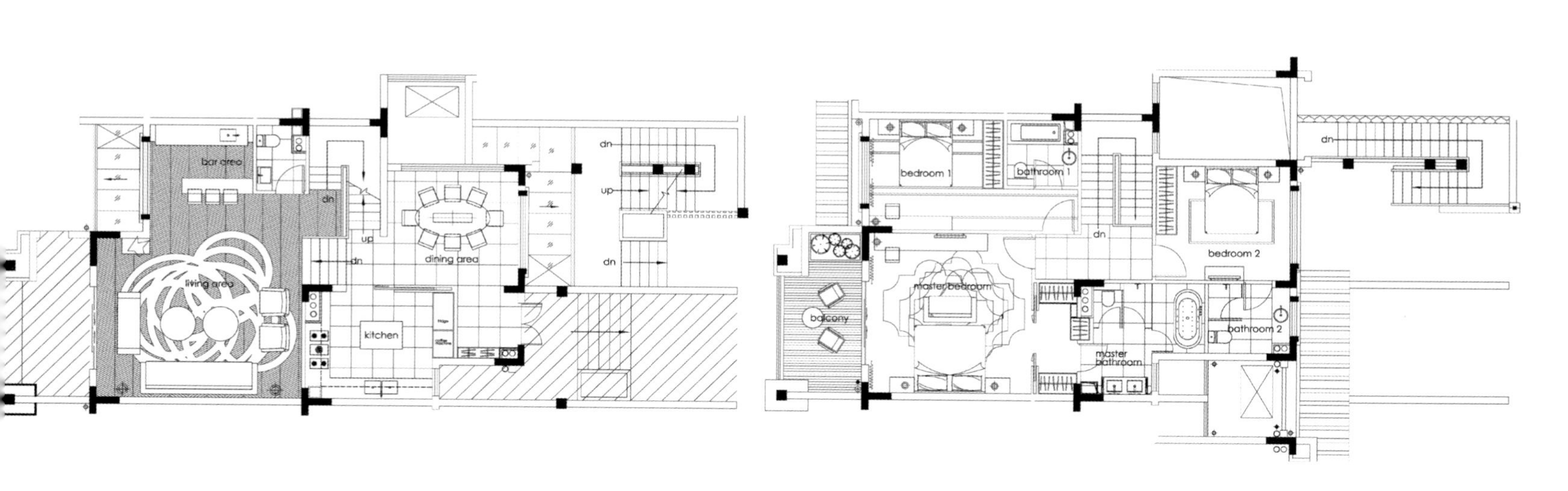
bar area
dn
up
dn
dining area
living area
kitchen
dn
up
dn
bedroom 1
bathroom 1
dn
bedroom 2
master bedroom
balcony
master bathroom
bathroom 2
dn

有固定壁灯及床头桌，这些设计，非常特别。设计师考虑到整幢别墅样板房布满立体花线，其他的家具都以简约线条为主，减少花俏或造型过于复杂的装潢影响整体设计效果。

至于厨房的设计，因为强烈线条感的花线设计已经非常独特，因此设计师选用了黑色，配合黑色的云石墙身，同时橱柜、台面亦以黑色为主色，凝造出非一般的型格。

This show house is inspired by "molding". Molding is one of the important elements in the continental style, but it becomes too tradition to use molding in the ceilings and walls nowadays. As the result, designers input new elements and spread the molding all over the whole show house. With this innovative idea, not only enrich the visual effects and its vitality, but also enhancing the modern and grand style as well.

As the molding lines effect is strong, only white, black and red being selected as the theme colour. Living and dining areas are in white color, in addition to the white marble, there are molding in various length and sizes on the ceiling and walls. The molding being located seems randomly but well placed and

enhancing the spaciousness. Furthermore, to interpret the simple but stylish design, designers make use of the molding to become the lamp shade. Stairs handles and the unique furniture are well interpreting the western interior design style.

In addition, there are custom made screens in the dining area. To replace the curtain, designers make use of this white and see through screens, with the repeating molding pattern, to enrich the comfortable and relaxing atmosphere.

Another highlighted sight point is the feature wall of the master bedroom. It is made up of ample laser cut wood panels, piled up to form a three dimensional molding panel with built-in wall lamp and bedside table. To ensure this extraordinary molding panel being the chief attraction and avoid fancy décor affecting the overall design, the furniture is in simple shape and design details.

Similar to the living and dining areas, with the strong molding lines features on the cabinets and countertop, designers simply use the black marble to match with the theme color for the kitchen. Without using too much color scheme and complicated design details, this black color kitchen is really in cool style.

Project Name: The Parc Four Seasons- Show House (F2), Shanghai, China
Design Firm:One Plus Partnership Limited
Designers Involved:Mr. Ajax Law Ling Kit & Ms. Virginia Lung
Type:Show House
Photographer:Mr. Ajax Law Ling Kit
Area:280 m^2
Materials:Carpet, curtains, sheer, roller blind, fabric, leather, stainless steel, stainless steel in powder coating finished, mirror, glass, marble, corian stone, granite, plastic laminate, paint, tiles, mother of pearl, mosaic, timber floor, timber, veneer, wallpaper

项目名称：中国上海晶苑四季御庭 F2 户型别墅样板房设计说明
设计公司：壹正企划有限公司
设计师：罗灵杰、龙慧祺
摄影：罗灵杰
类别：别墅样板房
面积 :280 m^2
主要用材：地毯、窗帘、窗纱、卷帘、绒布、扪布、扪皮、镜钢、砂钢、不锈钢、玻璃、镜、夹纱玻璃、云石、仿石、麻石、防火板、墙纸、实木、木地板、木皮、马赛克、贝母、磁砖、油漆、 墙纸

行政总厨的天马行空
SOARING ACROSS THE SKY

这是一间充分展现屋主生活习惯及共对美食爱好的样板屋，设计师以“行政总厨”为题，表现出一个热爱厨艺的人总希望为身边的家人及朋友随时大展身手，烹调美味食材与众分享，因此通过室内设计将兴趣渗入家居设计当中，设计师巧妙地以不同的食材、厨具及烹调配件等为题材并将之演化成家具及室内设计，诠释了爱厨人士的喜好与生活空间这两者之间的微妙关系。

空间装饰除了设计创意的展现外，同时兼顾到实用功能性，除了不同食材、动物造型的家具外，为满足爱厨屋主的实际烹调需要而添置了大量专业配备，包括红外线无烟烧烤炉、电炸炉等这些通常在餐厅才会有的厨房设备； 同时在样板屋中配备专业厨房、四季食材储藏区，仿如在家中也能化身为专业厨师一样。

设计亮点除了上述的专业厨房设备外，从各处可以发现各种由动物、食物为概念而演变成的家具，这亦是其特色。进入屋子的第一印象，从首层中各样摆设饰件及家具便能令人一眼看出主人的兴趣。

从进入客厅开始，已经看到用镜钢造成一块块 “砧板”形状的茶几、以餐碟堆砌而成的角几，以及地毯亦有如牛的形状的家具设计，另外还有饭厅，用上仿如面条造成的灯具，令整个首层充满了玩味及特色，成功让人融入这个以烹饪及美食所营造的气氛。

二层的水吧区，用叉子掘成椅子，加上 “吧椅” 为圆形，感觉似一只只绵羊，在抽象中体现主体；睡房吊灯用奶樽型的玻璃樽造成，书 形状似棉花糖；而洗手间也精妙地以“蒸蛋器”造型设计成洗手盆。

主人房间，衣柜造型与旧式鞋柜相似，配有大型金属铰及抽手。墙身的特色墙用上一些叉子，幻化成飞镖般随意粘贴于墙面，营造形式别致的画面及带出不一样的惊喜玩味效果。另外，模仿旧式鞋柜及自助餐炉的形态所造成的衣柜与梳妆台，亦突

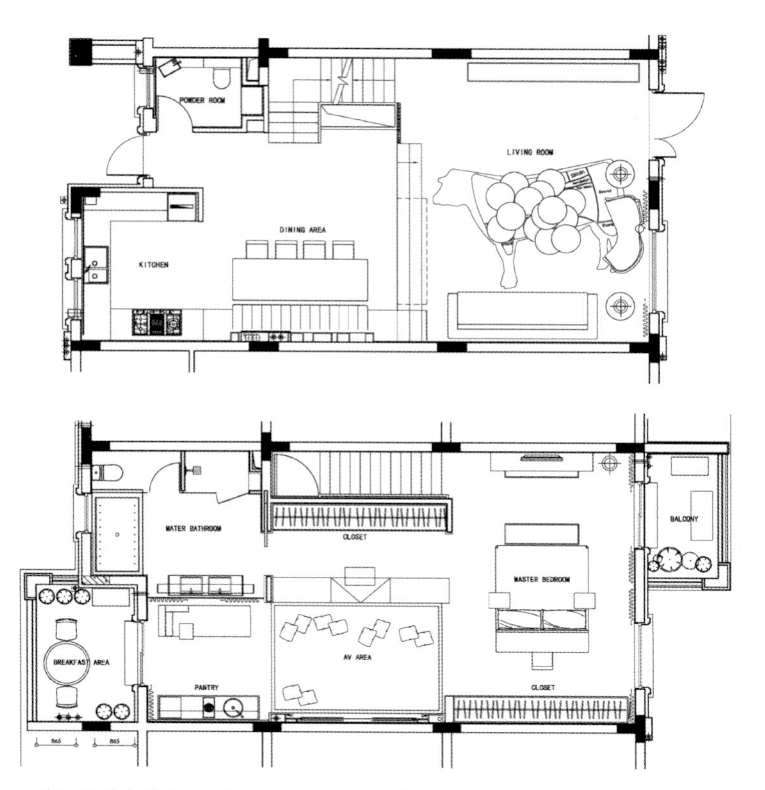

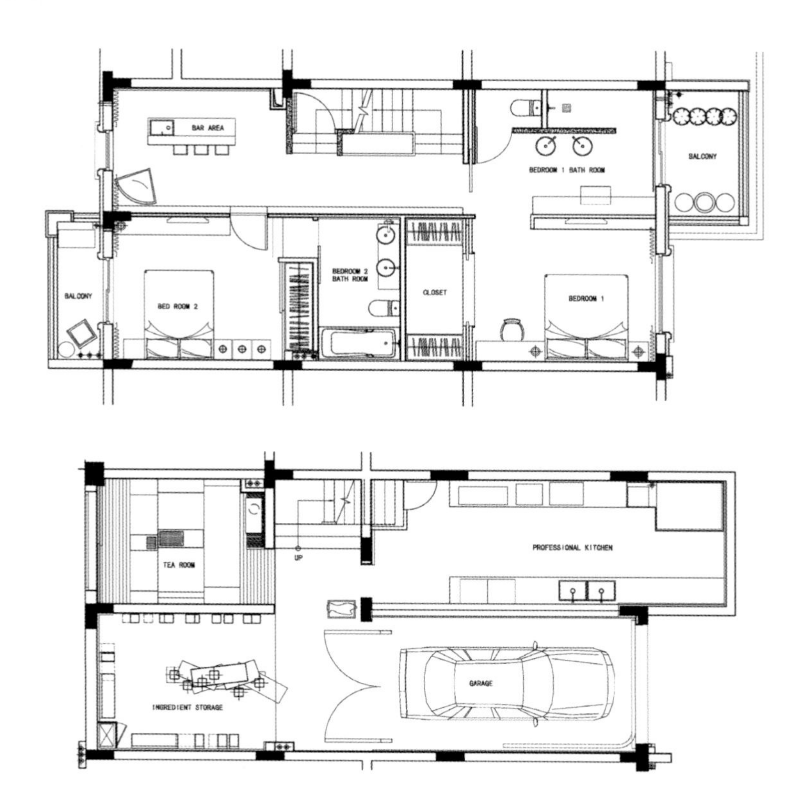

破了惯常家具的框框形态，创意中更见格调高贵，十分独特。

设计师进行设计时有很多天马行空的概念，非常重视新颖及创意的元素，这间样板屋以热爱厨艺者的家为主题，由于烹调手法及食材的选择是可以无限变化的，因此在设计时每间房间都塑造出不同形式的烹饪概念。这不仅免却过于沉闷的感觉，亦充分展现出设计手法的多元化。

This is a show house using "Executive Chef" as the main concept and showing house owner' s cooking interests and food preferences. To visualize the concept, designers made use of various cooking ingredients and tools and evolved them into furniture and interior design in which can fully interpret the relationship between the living space and the living style of the house owner altogether.

In addition to the design creativity, designers considered the practical functionality as well. Not only the ingredients and animal shaped furniture, there are a lot of professional cooking equipments, like the infrared smokeless barbecue oven, electric fry furnace, etc. and there is seasonal ingredients storage area, the owner can cook the delicious meals for the families and friends with the full and well equipped settings even staying at home.

To give the first impression into the house, there are various kinds of animals shaped furniture on the first floor and demonstrating the owner' s interest obviously. In the living room, there are stainless steel "chopping boards"

stacking to form the coffee table, "plates" piling up to form the table and "cow" shaped area rug; while in the dining area, there are "noodles-liked" pendent lamp, the whole first floor is interesting. Furthermore, in the bar area, forks were beat to form a side chair and there are round "sheep-liked" bar stools. The pendent lamps are made up of milk bottles, the study table is in "marshmallows" shaped and the wash basins are inspired by the shape of the "egg cup" as well. In the master bedroom, the wardrobe is similar to the old style refrigerator with the big size metal hinges and pumping hands and the dressing table is using the idea of the food warmer using on buffet table. In the master bedroom, designers make use of forks and plastering, and turn these forks into interesting feature wall.

As designers want to show the living style of the house owner and visualize the theme of "in love with cooking", the furniture, artwork, lighting, etc. with all sort of fun shapes being designed and enhances the creativity and imaginative of the house.

Porject name: Forest House in Taipei
Design company: AX Design Group
Designer; Huang Yujie, Lin Jiaqing
Area: 93 m^2
Material: piano stoving varnish, artificial stone, spray painting

项目名称：台北林宅
设计公司：大器联合室内装修设计有限公司
AX Design Group
设计师：黄裕杰、林家庆
面积：93 m^2
主要材质：钢琴烤漆、人造石、喷漆

闪电的灵感，源源注入正生活能量
FLASH OF INSPIRATION: POURING ENERGY INTO POSITIVE LIFE

本案在设计之初，业主提出了三点要求：

1. 打造出具独特性、能符合业主个人特质的居住空间。
2. 在独特造形下仍能满足一切生活需求与机能。
3. 在白色调的原则下，营造出不生硬冷酷，有热情活力的居住环境。

该案业主要求即要体现个性又很实用，由此可见业主是一位理性与感性兼具的人。

设计师非常乐意为业主达成这个都市中的居住梦想。在与业主沟通的过程中，设计师感受到业主身上总是散发着一股强烈的正向能量，感染周围的人。在设计师看来：能量是一种可以影响人的心灵但却又很无形的存在。而空间的灵魂来自于业主，它应该是主人型格的一种表征。设计师以其性格及生活态度为出发点，企图在空间中呈现这样的意象：以闪电的形象作为设计灵感的出发点。

空间内主要由两大片板块组构而成，一片是横断在空间对角线的主体墙面，一片是漂浮于上方的天花板体，糅合梁柱位置顺势而下赋予其空间特性，两片板块相互产生了形体上的变化，转折、扭曲、脱开、翻转、连接……有如生命体般在动态中形成空间的面貌，自然巧妙地隐藏起屋子中央巨大柱体及横在客厅正上方的梁。同时一贯地连接起厨房，客厅到书房，延展出空间最大景深。另外在材质的选择上，也试图将不同个性的白反映出不一样的风貌，钢琴烤漆的主体墙面反射出生活在里面的人们的生活动态；在天气好时，地面白色抛光砖会将外面的蓝天白云悄悄地带入室内；以不同倾斜面构成的天花板，更让自然光线的投射有了最好的表演舞台。

一个彰显业主个性强力的空间就这样完成了。

VAIO

At the beginning of planning the case, the owner proposed three requirements that the house should has personality as well as utility. From which we could see the owner is a person who is both emotional and rational.

1.Creat a living space which should be unique and meet the personal qualities of the onwer.

2.It could meet all the life requirements and functions under the condition of being unique.

3.Creat a living space which is not stiff, cold but enthusiastic and dynamic, uncder the principle of using the white tone.

Designers are very pleased to realise the urban living dream of the owner. During the communication with him, designers felt a strong postive energy that always exuded by the owner and infected the people around. The designers believe that energy is a kind of presence which could influence a

VAIO

person' s body and soul however it is invisible. Well, for the soul of house comes from the owner, so it is a symbol of owner' s character. As a result, designers started the plan from the owner' s character and attitude to life, and attempted to show these in the space. The images of lighting is as a start point of inspiration.

The space is mainly made of two big plates. One of them is crossing the main well of space diagonal. The other is the one which is floating on the ceiling above which embracing the beam so as to creat the space character. These two plates happen to creat a physical change: transition, twist, turing off, turing, connection...which seems a life-like body formed in a dynamic space. It also cleverly hides the huge column which across through the beam above the living room. At the same time, it connects the kitchen, the living room and the study to extend the maximum depth of field space. Besides, about the material choice, the designers tried to reflect different personalities with different kinds of white. Piano painted main wall reflects the lifes of people who lives inside; while in good weather, the white polished briks on the ground will quietly bring the bule sky and white clouds into the house. Ceiling which is created by different inclined planes leaving the natural light its best performance stage.

Thus completed a living space that reflects the strong personality of the owner.

Project Name: Residential Dome in Oderzo
Design Company: Arch. Simone Micheli
Photographer: Juergen Eheim
Size: 80 m²

项目名称：Oderzo 住宅圆顶
设计公司：西蒙内米凯利建筑设计公司（直译）
摄影师：克林斯曼
面积：80 m²

一体成型的 3D 线条

3D LINES

颠覆传统性屋顶框架，开启别样家居生活。底层开阔空间，引领无限视线。水晶缎面镜材，饰以枝干纹理，围合工作室空间，予人以惊艳，但又坦然之印象。雪白的纯洁气韵，以橡木地板之温润作为铺垫。过防盗门，客厅空间开扬，内有沙发，简约、金属质感、棕色、直角。其头靠陡然向上，铺就不同寻常路，内置 LED 照明，伴有小型音箱。客厅、厨房，两方世界，灯光闪闪，写下同样温暖。餐桌透明玻璃材质，不锈钢腿支撑，升华空间纯洁气韵。

房间铺陈和谐统一，真乃别样宁静，别样情。另有一小型工作室，毗邻厨房，以清玻璃围合而成。白色在此更加炫目耀眼。健身房设计简约，一应俱全，以缎面玻璃幕墙自然界定。点射灯光，光源镶嵌于地板之上，光束万种，激起千万想象。楼上主卧，一床自领空间风骚。其光熠熠，其神闪亮。一旦入内，唯我独尊。一墙大型开窗，覆以白而透明硬纱。纱顶又有不锈钢片，彰显理性纯洁。一墙饰以镜材，延续下层空间气度。其后卫浴，墙体、地板，别有一番天地。水槽硬性表面，储物柜镜面涂层，空间优化至极至圣。其质感现代艺术，形体纯洁、神圣。沐浴的是身心，宁静的是灵魂。

卫浴是白色亮光漆面家具。其墙体同样缎面玻璃镜材装饰。镜后置以背光设施。各背光形像几何，高居于水槽之上，并与镜中自然成像。门户，家具把手不锈钢制。组合家具气势雄伟，不仅连接厨房、客厅不同功能的空间，而且一路宛转，直至二楼，俨然成为卧室墙体。其中一面，独为收纳式家具，怡眼、实用。

On the top floor of a residential dome in Oderzo, rises a masterpiece with essential frames that strongly refuses any conventional rule introducing a new personal living. The stereotype transcends from the first moment the visitor enters this world. The ground floor is depicted to the eye as an open space, a huge studio apartment separated by crystal clear huge sheets of satin finish glass with stylized tree trunks design astounding yet calming the welcome guest. A snow white environment covered with oak parquet floors. Over passing the security door a wide space with a simple metallic brown angular sofa whose headrest rises all the way up to the ceiling, undertaking an irregular but functional path. Along this soaring snake, small luminous led reveal the incorporated speakers, brightening the living room and in the same occasion the adjacent kitchen. Below the sofa, RGB multi color LED changes the atmosphere of the lived space. The adjacent kitchen and its large dining room are presented in a shiny lacquered white and seem hosted in the living room while the dining table is made of transparent glass supported by stainless steel legs. The purity that characterizes the environment spreads all around the room, investing the furniture unit that is also the sofa frontal wall on which a TV illuminated by RGB led is mounted.

Everything is in complete coherence and harmony with all the rest conveying a sense of serenity to the human visitor. Adjacent to the kitchen one can glimpse at the tidy studio through the clear glass wall. Here, the room is also characterized by a dazzling white color; three revolving libraries with rear mirrors allow you to change this area to your liking; another transparent satin finish wall separates this space from a simple and essential gym. Each of the anthropomorphic geometry stylized on the transparent wall is enhanced by technical spotlights whose light rises from the floor, highlighting the imagination. There is a strong euphony between the white which apparently characterizes almost the entire apartment, the mirrors and transparent satin finish windows glazing all over the ground floor.

Immediately after, we find the bathroom with glossy white lacquered furniture, and again, the satin mirror glass plays an important role in the decoration, filling almost all the walls. The satin finish mirror with rear lights and its natural geometric figures in the background appear important above the sink,

differently from all the other environments, they are mirrored. Each handle of the house is made of stainless steel, from the security door to the giant furniture, which joins the kitchen, the living room and reaches all the way to the upper floor becoming one of the bedroom walls. The storage furniture covers an entire wall of the house appearing very pleasant, but also very useful.

The mansard bedroom on the upper floor is presented with one bed fermenting in a luminous glowing color, centering the person who will rest and sleep on it. A white organza curtain embraces the enormous window, showing a veil of stainless steel on top, while a mirror similar to the ones on the lower floor covers the opposite wall. The private bathroom is concealed here, different from all the rest, with walls and floor in green labradorite. The sink is in Solid Surface, while a storage cabinet externally coated with mirror was created to optimize space. This contemporary work of art proves to be a candid and composed environment, distinguished by pure and hieratic geometries; a comfortable and meditative cradle whose atmosphere connects the visual with the interior harmony, mirror of our soul which becomes more serene and balanced.

Project Name: Elena and Marco Bergamo House
Design Company: Arch. Simone Micheli
Photographer: Juergen Eheim
Size: 300 m²

项目名称：贝加莫楼
设计公司：西蒙内米凯利建筑设计公司（直译）
摄影师：克林斯曼
面积：300 m²

自由流淌的创意之家
THE HOME OF ORIGINALITY

本案空间贝加莫楼，位于顶层，其量体形状奇特，极具视觉冲击。但内里设计，简约纯洁。梦幻气质，现实生活，彰显家居之温馨。优化空间，简单铺陈，都市喧嚣中难得如此清静归隐之处。

餐厅中别有洞天。长方形映像中，正中一餐桌，钢轴支持。客厅、餐厅、厨房以其为界，并彼此相互连接，足见设计构思之精巧。白色的马赛克装饰，缠绕于银器之中，那是现代的古韵。古韵之中，另有平台，以彩灯为饰。背灯电视卡座，镜面框，简约灰色布艺直角沙发立于其上。过门空间层次分明，上是迷宫。其中镜面林立，置以背灯，小小的显示器位于其中，彰显映像之神秘、百媚。下是两个柜组，亮光白色漆清。其白如纯，呼应空间墙体、天花之同色系。右面出一立体装饰。细打量，其后一简约客用卫，以镜材无缝对接。小小的空间，因此纵深延展。

极象征之意象，精华之根本是本案空间的气度。或纯洁，或简约，但却别致。并不宽阔的空间，却被大师以强烈、敏感手笔，全心打造梦想生活。

长长的走廊以玻璃推拉门开启。其深如巷，星星点灯，那是黑暗中的一抹温柔。走廊右侧即是卧室，左面三扇门户，引领卫浴空间。其内以镜子马赛克装饰，并有健身空间。亲密之气氛直泄卧室。卧室白色床头。床头镶以镜材。镜材银光与各壁尾灯自成呵成之势。推拉门旁，超大橱柜，正对卧室。旁还有一私人浴室，内以多功能土耳其淋浴设备装饰。浴室之外，大面缎面磨光镜面沐浴在光华中。在这种光韵中，圆柱形洗手台气势恢宏，但悬在空中，呈飘渺轻盈之态。

The Elena and Marco Bergamo House in Oderzo is portrayed as an irregular purity, with unconventional shapes, a spacious rebuilt penthouse in a residential house, redefined as a simple and outstanding masterpiece of men' s life. This multi sensory architecture work carries the visitor towards a dreamlike yet safe dimension, where dreaming becomes a revealing emotional experience. The space optimization and the furniture simplicity discloses the masterpiece anthropocentric' s vision to leave out some free space for the primordial exercises deeply eclipsed in our mind by the speedy and noisy world. The intention is to spread the balanced awareness that it is possible to create a rational and unique tranquillity capable to set aside any unbalance present in our dynamic daily life.

The accommodated framework in the dining room has a rectangular hole onto which the dining table supported by a steel shaft, also acts as a simple white parting between the living room, dining room and kitchen. White mosaic decorations with silver features tangling around each other. A TV carter with rear lights and a mirror frame leans on this framework in front of which there is a simple grey angular sofa in fabric pleasantly fermenting on a platform filled with colored lights. Walking past the door, on the wall behind the visitor' s shoulders there is a labyrinth made of mirrors with rear lights and small monitors that give way to the most mysterious visual fantasies; immediately below, there are two cabinets in glossy white lacquered varnish. White is the absolute protagonist in this room, covering every wall and the vault. The walls are coated with lacquered MDF wood and host a multitude of cabinets with stainless steel handles. On the right, a protrusion hides a simple and essential guest bathroom with a jointed game of mirrors that gives depth to this small environment.

Adjoining, across a glass sliding door one enters a long corridor illuminated by small lights that escort the intimate atmosphere in the dark. On the right of the corridor we find the bedroom while there are another three doors on the left, giving access to a bathroom decorated with a mirror mosaic, and a gym. Intimacy also invades the bedroom. A bed leans on a white headboard frame, onto which a long mirror is fitted with rear lights that runs almost along the entire wall. Facing the bedroom near the sliding door there is a huge cabinet behind which lies a private bathroom with a multifunction Turkish shower-bath. The large cylindrical wash basin is suspended on a dazzling colored light and positioned outside the bathroom in front of a large illuminated satin finish mirror.

The symbolic and essential architecture distinguishes this Oderzo apartment in an indisputable way. Purity and simplicity expand in the environment in order to capture, tame, and turn any kind of emotions into a unique and unavoidable experience. Such a strong and sensitive masterpiece attracts the user who searches deep in the most cramped spaces of his soul, dreaming to live his own life in "another" way.

Project Name: Alpine Villa
Size: 1 200 m²

项目名称：阿尔卑斯别墅
面积: 1 200 m²

旋转出生活的无限灵感
INSPIRATION IN ROWS

阿尔卑斯山，布莱德湖成就着座座与众不同的别墅。本案建于19世纪，依山傍湖，风景秀丽。业主希望改建后的建筑空间主生活区域能扩大一倍。静坐于各功能空间， 淡看湖上风云。为此，诸建筑师特邀而来，使其别墅成为集思广益的成果。

但700 m² 的新增空间，如何融入原有建筑量体，同时不对原有自然环境造成压力，这却是不大不小的难题。

经过设计师的妙手，置新增空间为地下掩体，无不是精彩的一笔。

空间中圆形的地基令人遐想不尽，如锦缎、丝绸般的枕头，而周围的景观如同枕套，锦上添花。地下空间流光溢彩，湖景山色，尽收囊中。地表高度分层明晰，50 cm 的落差，展现着空间丰富的表情。平台位居枕上。顶层花园，书意绿意，是孩子们娱乐、嬉戏的所在。

主生活空间，如配套厨房、餐饮区、视听室、壁炉房、图书馆、工作室等皆位于新添的量体内，很好地保持了原有建筑的布局与景观。建筑传承的理念顷刻间得到了体现。

日常服务设计，如衣柜、主厨房、客用浴室，分别位于原酒窖之后，车库的存储间变为隐性设计。原有的一楼和顶楼专为孩子、父母享用，孩子的天性、老人的静养得到了人性化的充分关怀。

大门内庭院几许。入口直连房间中轴，一面朝向新添加量体，一面通向坡道，直入楼梯。大厅三层的空间，旧时恢宏的气质依然。不觉时，已然身处于临湖的生活起居空间。

楼梯攀援直上，各功能空间及主厅依次排列，俨然整个空间的核心所在，同时划分着新旧两个空间，形成相融相生的两个世界。

The project involves an extension of a 19th-century villa located in a beautiful Alpine resort next to Lake Bled. Both the old villa and the landscape were strictly regulated by the National Heritage people. The client' s chief request was for the main living area to be twice the size of the old existing villa. On top of that, most of the spaces had to face the lake. The client invited several architects to come up with ideas.

The main task was how to incorporate the new 700 m² addition while respecting all the restrictions the villa and the surrounding landscape were subject to.

Our proposal placed new spaces beneath the ground floor of the existing villa.

The extension forms a rounded base around the house – a pillow covered by the landscape. Looking from the other side of the lake, the pillow blurs with the surrounding landscape. The elevation under the pillow is glazed and overlooks the lake. The floor is organised in levels according to the outside landscape (±

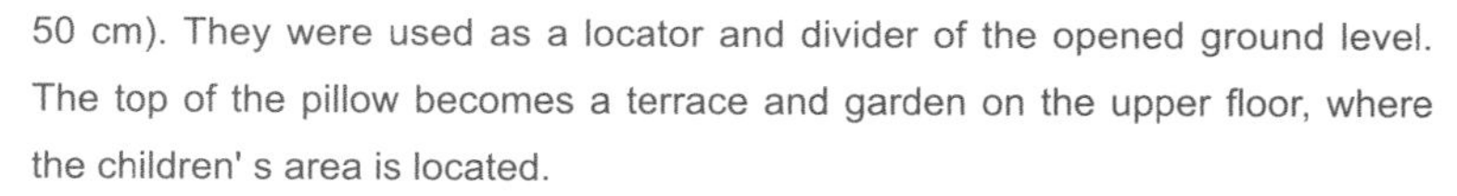

50 cm). They were used as a locator and divider of the opened ground level. The top of the pillow becomes a terrace and garden on the upper floor, where the children' s area is located.

The main living spaces (the ancillary kitchen, dining area, TV space, a space with a fireplace for listening to music by, a library and work space) are placed in the new extension.

Everyday services such as wardrobes, the main kitchen and guest bathroom are behind the walls of the old villa cellar. Garages and storage/workshop spaces are hidden in this volume. The old villa becomes a private rest area with the children' s space on the first floor and the parent' s on the top.

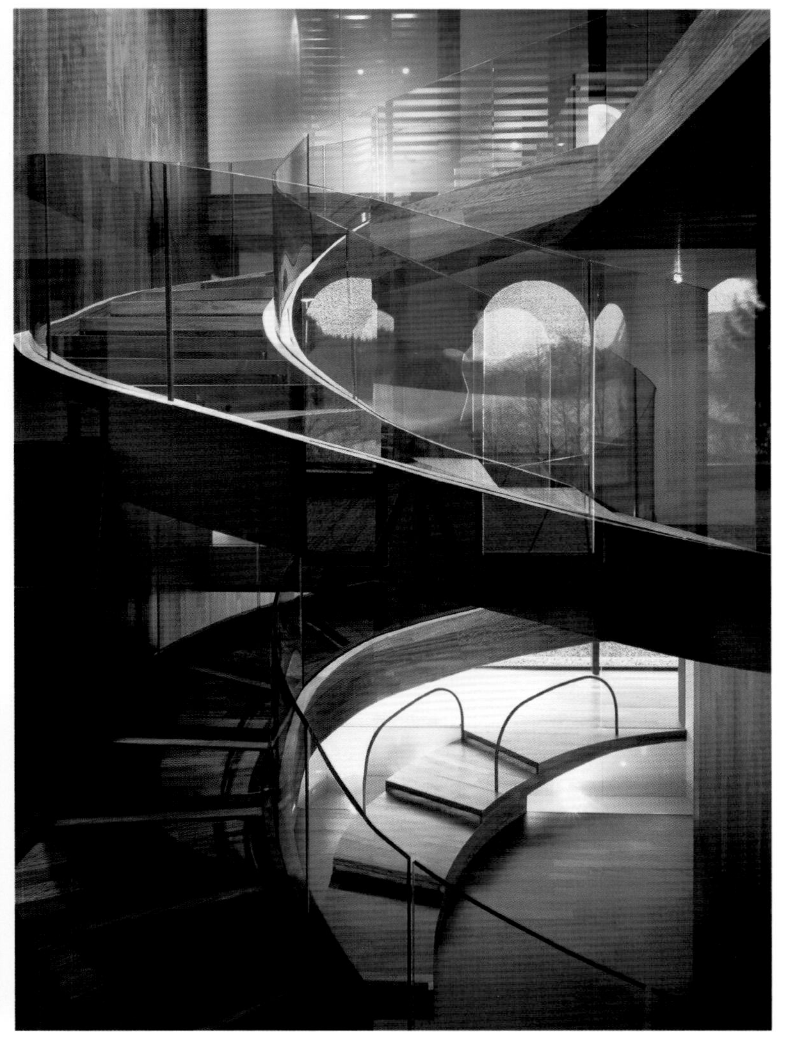

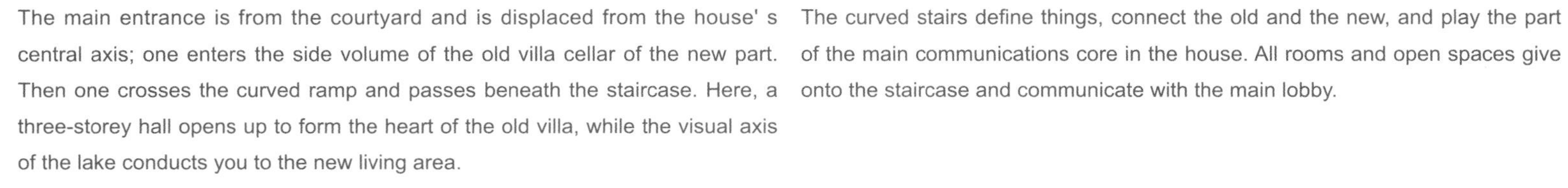

The main entrance is from the courtyard and is displaced from the house' s central axis; one enters the side volume of the old villa cellar of the new part. Then one crosses the curved ramp and passes beneath the staircase. Here, a three-storey hall opens up to form the heart of the old villa, while the visual axis of the lake conducts you to the new living area.

The curved stairs define things, connect the old and the new, and play the part of the main communications core in the house. All rooms and open spaces give onto the staircase and communicate with the main lobby.

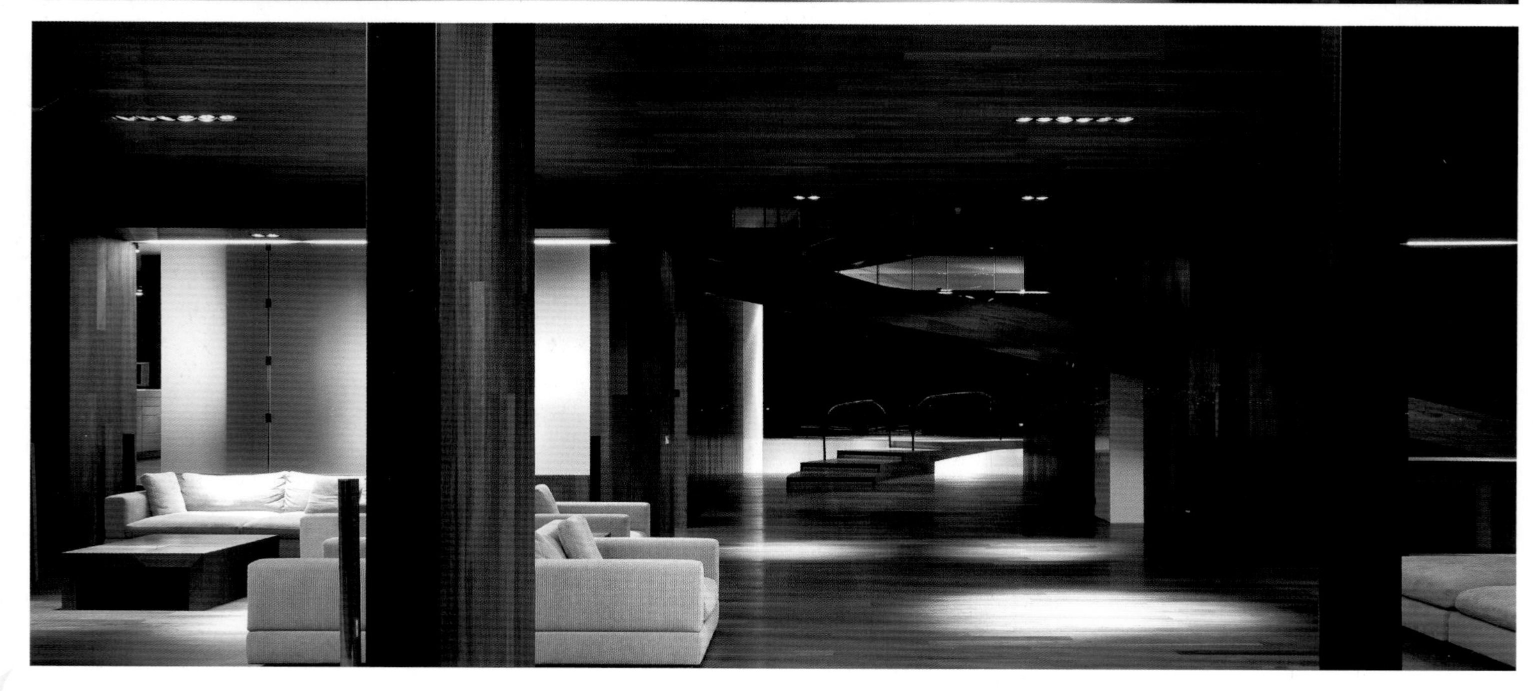

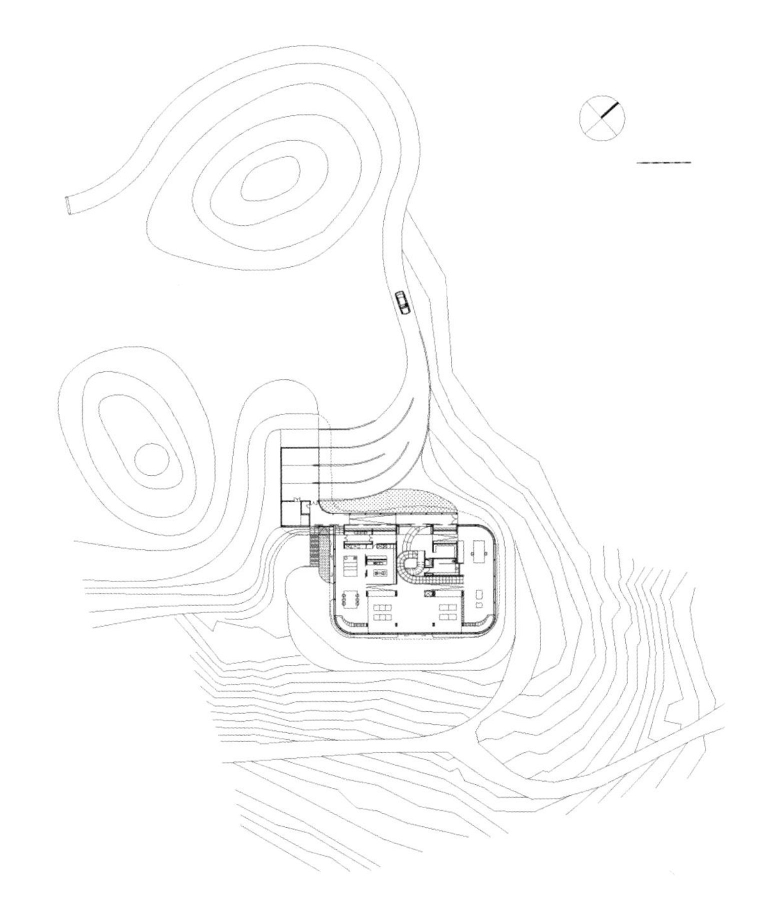

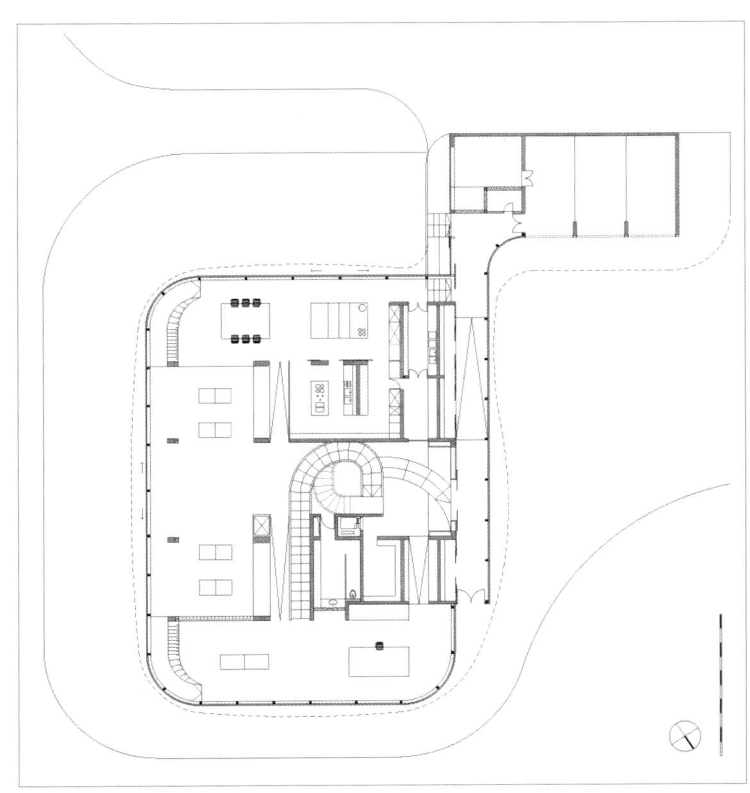

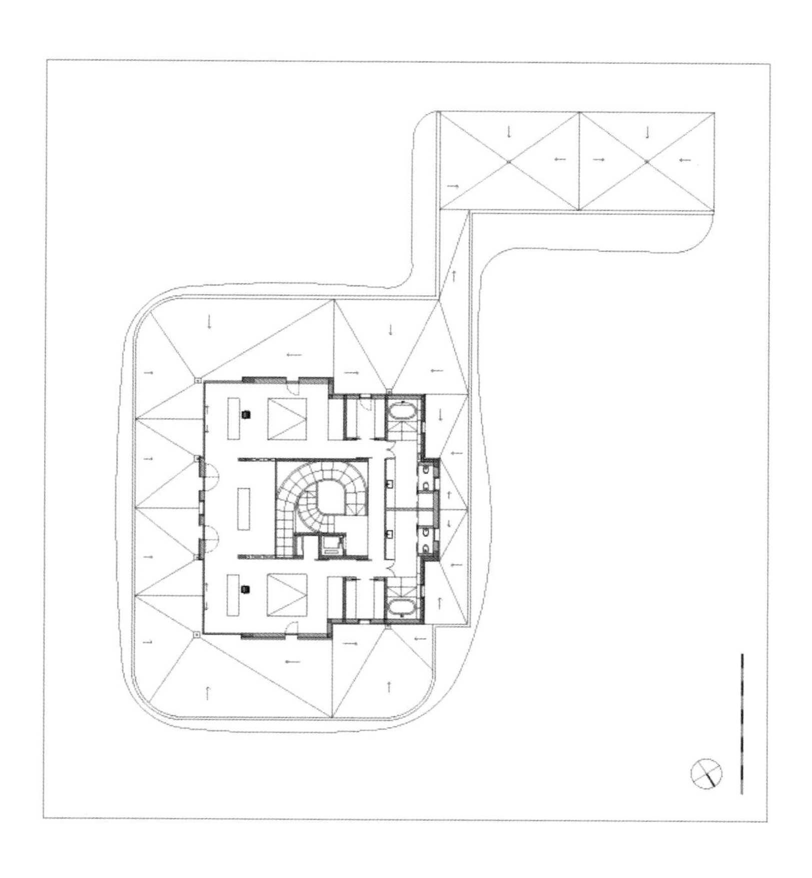

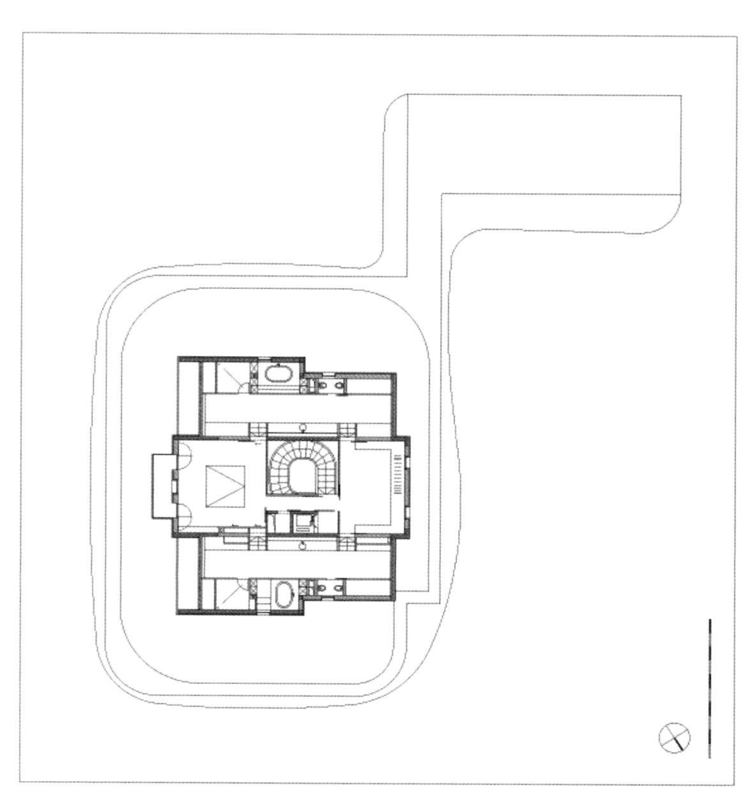

Project Name: Villa MN
Design Company: UN
StudioDesigner: Ben Van
Berkel Size: 300 m^2
Materials: Reflective Glass

项目名称：MN 别墅
设计公司：UN 工作室
设计师：波克利
面积：300 m^2
用材：反光玻璃

空间与自然的扭转拼接

THE SEAMLESS STITCHING OF SPACE AND NATURE

倾斜的基地似乎专为本案单体别墅而生。两个相互独立的量体，一无缝对接于北坡，一陡升于南巅。一高一低，一南一北，相互呼应，共成盒状结构。南巅之下，内有车库；五平行墙体，沿水平轴线，从水平至垂直划分各功能空间。各空间错落有致，尽成立体、三维之感。宽大的开窗，地板，天花，室内，室外尽成一体，如潺潺流水，过山川，越林木。

In the design for this single family house the sloping site is used as a device for programmatic and volumetric organization. A box-like volume bifurcates into two separate volumes; one seamlessly following the northern slope; the other lifted above the hill creating a covered parking space and generating a split-level internal organization. The volumetric transition is generated by a set of five parallel walls that rotate along a horizontal axis from vertical to horizontal. The ruled surface maintaining this transition is repeated five times in the building. From inside the huge window strips from floor to ceiling allow a fluid continuity between interior and landscape.

project name:Arc House
design Company: MB Architecture
Photographs: Courtesy of Maziar Behrooz Architecture, Matthew Carbone

项目名称：Arc 住宅
设计公司：MB 建筑设计公司
摄影：Behrooz 建筑公司 马修·卡蓬

山坡上的圆顶屋
DOMES ON HILLSIDES

这间大理石打造的拱门造型房子坐落于昌迪加尔城市周边 8.7 km 的地方，该项目作为昌迪加尔新型住房的蓝本，目的是想建成可持续生态的新型住宅。设计师不负众望，该项目空间配置不但解决了可居性问题，同时协调了环境与社会问题。

项目由来自南开普敦的 MB 设计公司负责设计，空间的规划主要是通过隔离边缘所有外界喧嚣，从而在中心创造一个环保静谧的居住环境。

弧形屋顶使项目看起来更为独特，带有无限美感。内里空间设有一个大堂（也即我们传统意义上的大厅）以显示一种社会感，其他生活空间亦将普通的奢华理念及简约、传统空间完美结合；而住宅外部主要使用不同的材料进行弧形打造，顶层起到了阁楼的作用，远看犹似格栏的叠加，不仅提高了空间的功能性，也使得住宅空间和周围景观从步行区到绿化带的连接得到细分。设计师考虑到该项目只是低层的新型住宅，还在每一层提供可以和地面有联系的地台，使其环境得到充分利用。

Marble Arch is a housing development located in Chandigarh on a 5.4 acre site along the periphery of the city. The project' s objective is to develop a new prototype for housing in Chandigarh as an entity to address issues of liveability, spatial configuration, environmental and social issues, while shifting away from the archetypal morphology of high specification residential modules and equipment crammed into an undersized apartment.

The luxurious, sustainable home was designed by MB Architecture, an architecture firm from East Hampton. They combined the luxurious value of The Hampton-associated mansion and the simplicity, traditional aspect of The Long Island one. With its arc roof, this sustainable building looks very unique, with stunning aesthetic.

Each block within the development has an atrium lobby to provide a sense of community. Given the fact that this is a low-rise development, the opportunity to provide terraces on each level to be able to establish a relationship with the ground level has been fully utilized.

The outward expression of the housing relies on a varied use of materials which are carefully chosen to enhance the individuality of the spaces within. The design employs the use of grids being superimposed on the entire scheme both in the case of buildings, where it gets subdivided to generate spaces within the apartments as well as onto the landscape by way of pedestrian linkages and green areas.

SIMPLE AND PURE
简素
以质取胜，清空杂念，放怀天地间
MEDITATION PRACTICE QUALITY-BASED, OPEN-MINDED
BETWEEN HEAVEN AND EARTH

Project Name: Naka House
Design Company: XTEN Design
Designer: Monika Haefelfinger, Austin Kelly
Materials: Venetian Plaster, Epoxy Resin Wood, Painted Metal

项目名称：仲楼
设计公司：XTEN 设计
设计师：莫妮卡、奥斯汀
用材：威尼斯石膏、环氧树脂木板、涂色金属

天籁般纯净的素色居庭

THE COURTYARD PURE AND HIERATIC

HOLLYWOOD

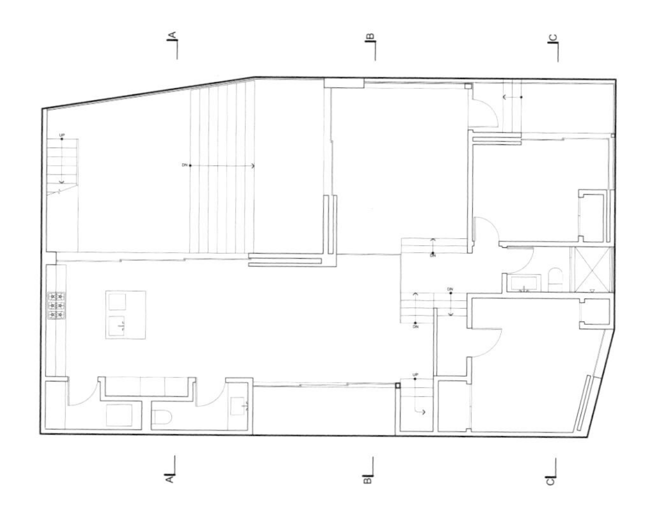

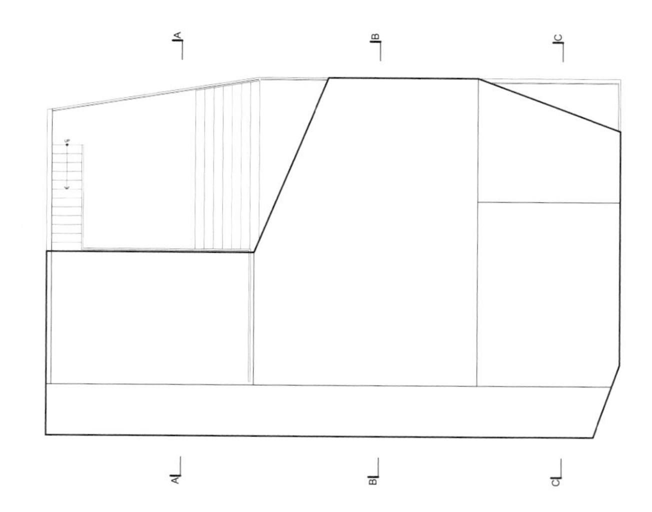

本案量体原建于 20 世纪 60 年代，骑于好莱坞山脊，水平略低于好莱坞标志。南、西两面俯瞰观毛榉树峡谷；东临自然保护山谷，尽观格里菲斯公园天文台。
建筑基地倾斜，高低层次不同。重新设计后的量体空间立足于现实，建筑现状，室内重新铺排，外部实现贯通，充分发挥山景、林木之天然优势。厨房、饭厅新加大型露台，一钢构楼梯连接屋顶平台。卧室侧翼，另加阳台。上层主卧，从天花至地板，滑动玻璃面板，界定与连接不同空间。
外墙圆润、光滑，黑色灰泥。绿色自然，生机无限，静静量体，雕塑质感。内里空间，环氧树脂地板、平台，漆面橱柜、金属板材等等。用材，各不尽同，但同样的框景手法设计，凸显空间物理优势。
内外对比强烈。外部山野景观，极其自然随意。内部井然有序，迥然相反，一旦入内，纯白色系，延伸的是空间，抽象的是气氛。

Naka house is an abstract remodel of a 1960's hillside home located on a West facing ridge in the Hollywood Hills, just below the Hollywood sign. To the South and West are views of the Beechwood Canyon; to the East is a protected natural ravine, with a view of Griffith Park Observatory in the distance.
The existing home was built as a series of interconnected terraced spaces on the down slope property. Due to geotechnical, zoning and budget constraints the foundations and building footprint were maintained in the current design. The interior was completely reconfigured however, and the exterior was opened up to the hillside views and the natural beauty of the surroundings. A large terrace was added to link the kitchen/ dining area with the living room, with a steel stair leading to a rooftop sundeck. Terraces were also added to the bedroom wing and the upper master bedroom suite to extend the interior spaces through floor to ceiling glass sliding panels that disappear into adjacent walls when open.
The exterior walls are finished in a smooth black Meoded ventetian plaster system, designed to render the building as a singular sculptural object set within the lush natural setting. A series of abstract indoor-outdoor spaces with framed views to nature are rendered in white surfaces of various materials and finishes; lacquered cabinetry, epoxy resin floors and decks and painted metal.
The contrast between the interior and exterior of the house is intentional and total. While the exterior is perceived as a specific finite and irregular object in the landscape the opposite occurs inside the building. Once inside the multitude of white surfaces blend the rooms together, extending ones sense of space and creating a heightened, abstract atmosphere from which to experience the varied forms of the hillside landscape.

Project Name: Loft Apartment
Design Company: 2B Group
Designer: Vyacheslav Balbek, Olga Bogdanova, Alena Makagon
Photographer: Vyacheslav Balbek
Size: 630 m^2

项目名称：三层别苑
设计公司：2G 集团
建筑师：维亚切斯拉、奥尔加、阿伦娜
摄影：维亚切斯拉
面积：630 m^2

生态与艺术的协奏曲

THE ARTISTIC ECO-CONCERTO

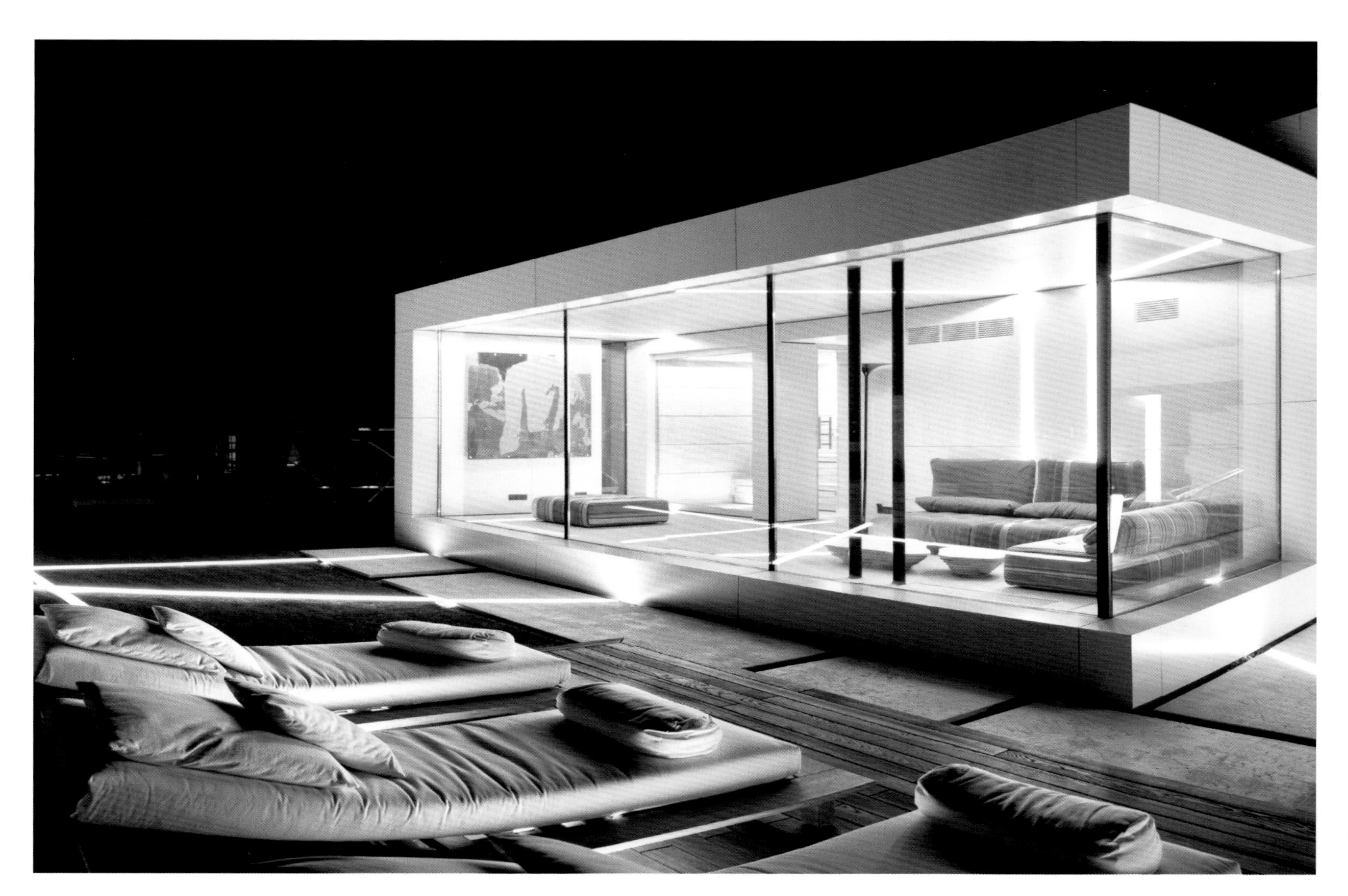

基辅中心区，第聂伯河河畔。本案为三层顶楼复式，380 m^2 的室内空间。东南面另有楼台，面积 250 m^2，扶栏凭望，独享基辅城市风光。

空间集设计之精华，质感、色调、造型、用材实现和谐对接。

木材、水泥、玻璃、金属、砖，尽在主生活空间。其内线条，细部出其不意地令人惊叹。绿色幕墙，自然之意象。混凝土酒架，盲文字样图案。阿特米德灯，高大阔气。两射灯，垂直光束，自是一地银光。

下层空间，绿墙延续。家用办公室，气氛轻松。次生活空间，引领主卧世界。睡眠区，另有一卫浴隐于移动木盒之内，有生活之便利，无视野之阻断。除些许红、绿氧化铜板隐于墙体之内，空间主打中性色调。

儿童房及其浴室，愉快大胆，色调明快。墙壁更有留白，专用于极兴来临时涂鸦之用。开放屋顶平台，以几何落地灯作为点缀。另有整块石料铸就的酒吧、壁炉、手工木板长椅、玻璃亭台。冬季酷寒，此处仍是温暖一片。

The project itself was a top floor 3-level penthouse with total space of 380 m^2, located in the heart of Kyiv on a Dnieper river. Third floor of the apartment comprised an open terrace facing Kyiv left bank panorama on a south-east with total space of 250 m^2.

The project is an interior that captures a multitude of styles, all designed to offer a seamless collection of textures, colors, shapes and materials.

Wood, concrete, glass, metal, exposed brick – all these materials found their place thighs main living area and surprising lines and details were shaped with their help. A green wall in the living-dining space brings nature indoors. Living area also features numerous details like the concrete wine rack with the Braille type dots. All ceiling light has been substituted with a giant Artemide lamp and two vertical spotlight projectors filling the area with reflected smooth illumination.

The green wall continues down to the lower floor, where a lounge-home office space and a second living space lead to the spacious master bedroom joined with a bath right behind a bed back. In order to hide the master bedroom toilets, authors invented a mobile wooden box stylized as a large parcel. All in neutral colors except for red and green oxidized copper sheets serving as hidden shelf facades.

The children' s room and bathroom look cheery and bold due to young artist paintings on walls and graffiti tags on concrete background.

An open rooftop terrace is lined with geometric floor lights and features monolith bar and a fireplace, handmade terrace plank benches and a glass pavilion for winter leisure.

DESIGNER
TOKYO
ARCHITECTURE INSPIRATIONS

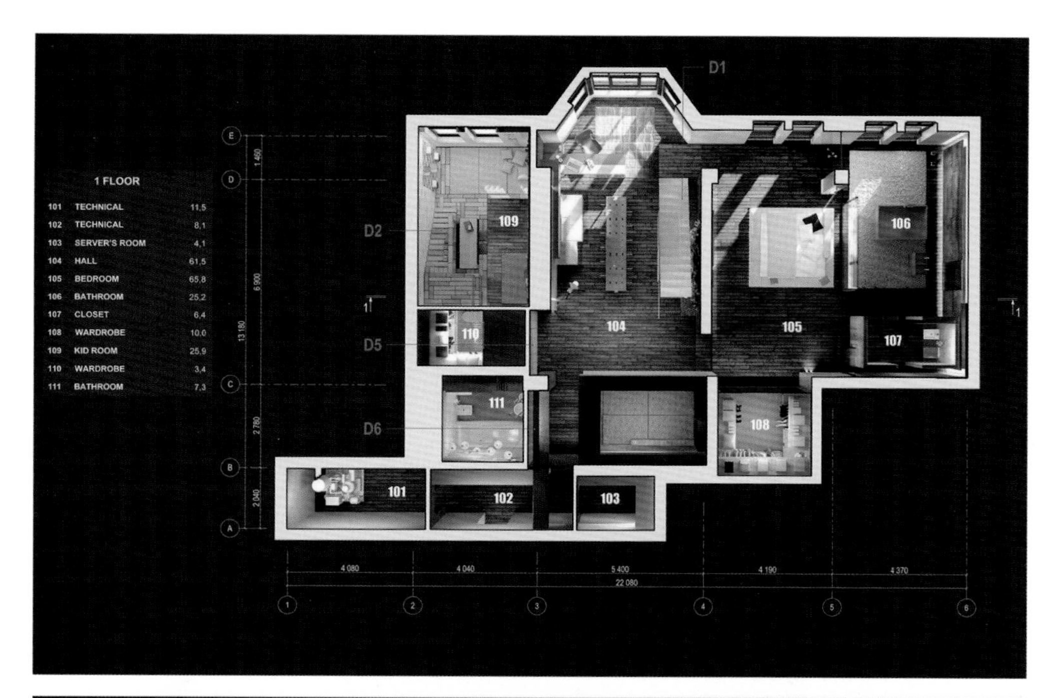
1 FLOOR
101 TECHNICAL 11,5
102 TECHNICAL 8,1
103 SERVER'S ROOM 4,1
104 HALL 61,5
105 BEDROOM 65,8
106 BATHROOM 25,2
107 CLOSET 6,4
108 WARDROBE 10,0
109 KID ROOM 25,9
110 WARDROBE 3,4
111 BATHROOM 7,3
D1
D2
D5
D6
101
102
103
104
105
106
107
108
109
110
111
4 080
4 040
5 400
4 190
4 370
22 080
1
2
3
4
5
6
E
D
C
B
A
1 460
6 900
13 180
2 780
2 040

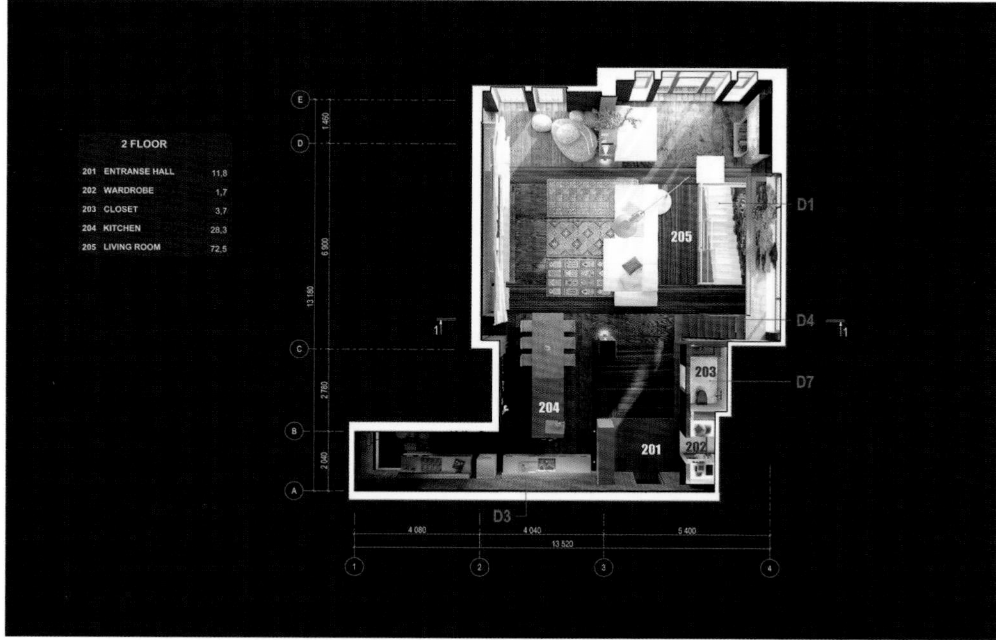
2 FLOOR
201 ENTRANSE HALL 11,8
202 WARDROBE 1,7
203 CLOSET 3,7
204 KITCHEN 28,3
205 LIVING ROOM 72,5
D1
D3
D4
D7
201
202
203
204
205
4 080
4 040
5 400
13 520
1
2
3
4
E
D
C
B
A
1 460
6 900
13 180
2 780
2 040

TERRACE
301 RESTROOM 18,4
302 INFRARED SAUNA 5,0
303 WC & SHOWER 2,6
304 ENTRANCE HALL 4,2
301
302
303
304
D8
1 460
6 900
2 780
2 040
5 100
18 280
E
D
C
B
A
A1
8
7
6
5
4
1 200
4 200
4 370
13 960
4 190

HE IS
THE ONE

13

HE IS
THE ONE

Project Name: Kempinski Bellevue Residence
Size: 210 m²

项目名称：伊斯坦布尔别苑
面积：210 m²

光影魅惑
LIGHT ENCHANTMENT

本案 210 m²，风景独好，享尽伊斯坦布尔之繁华。客厅、开放式厨房、主卧、步入式衣帽间及主浴各一。客卧、客用浴室各二。正对玄关，一衣架单独设立。玄关右侧，置一玻璃书架，内存古书，书香阵阵传出，自然过渡空间、彰显设计之理念。书之古朴，玻璃之现代，对比明显，飘逸在内里空间。厨房、客厅原本各分离在不同空间，如今合二为一。长长的卧榻，专为本案设计，家居怡然，尽在其中。主卧，原用以隔断卫浴、睡眠区的界定墙已无踪影，取而代之的是一扇玻璃幕墙、推拉门。左观、右看，尽是美丽自然景观。

空间用材，功能设计，建筑、设计元素，如墙壁、天花、地板，和谐融合于空间。家具、配饰适时点缀，为空间刚性气质添了一抹柔情，书写一笔平衡。

用材的选择，色度的选择延续平衡笔触。一边是天然石墙面、玻璃隔板、彩绘玻璃、绸缎抗锈表面，一边却是复古式实木复合地板。一边是熟悉、温暖、真实、经典的配饰，一边却是清爽、明快但看上去依然显几分冷峻的用材。

虽然色彩单一，但生活空间依然舒适休闲和优雅。这原本就是业主的期望。

照明设计，集功用、美观于一体。绘画、铺陈，以射灯点燃生命华彩。灯具闪亮，彰显明目，但更为耀眼的却是那一丝亮丽。可能，这才是光华的真正意义。

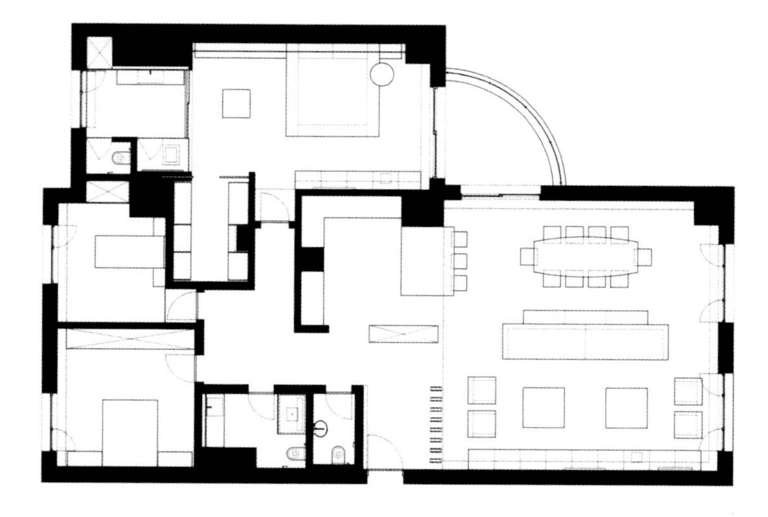

86.0 m^2

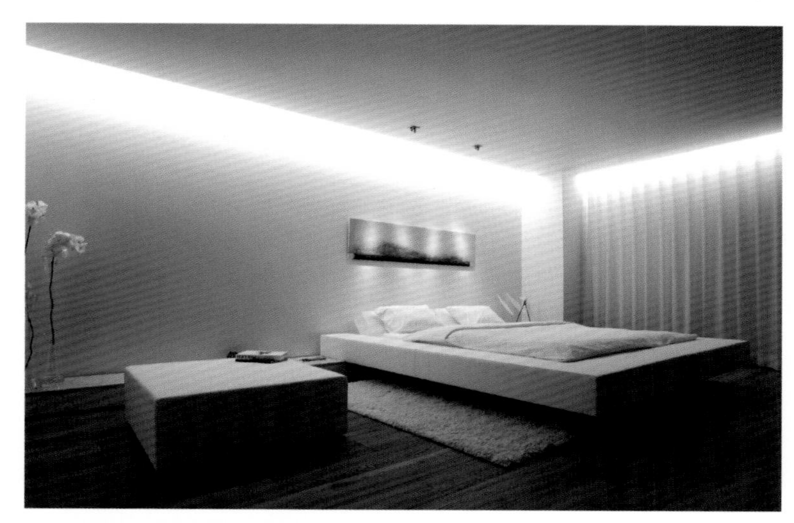

Kempinski Bellevue Residence is a 210 m^2 residential project located in the Bellevue Residences enjoying a panoramic view of Istanbul. The apartment is designed to feature one living room, an open kitchen, one bedroom with walk-in closet and en-suite bathroom, two guest bedrooms and two bathrooms. The coat rack, which is not included in the current plan, is located right opposite the entrance as a free standing object. On the right hand side of the entrance is placed a glass bookshelf with old books. This is also used as a separation between spaces and is reflecting the main concept of this project. The transparency of the material, the modern form and the elegant looks contrast with the traditional and classic appearance created by the old books, defining the main atmosphere of the house. The kitchen, which in the original plan was separate from the living room, is now open creating a larger living space. The long couches were exclusively designed for this project to enhance the comfort of the home. The special design furniture features an elegant character while offering sensual and physical relaxation. In the master bedroom, the separation wall between bedroom and bathroom was removed and replaced with a transparent glass partition and a door. This guarantees the view and natural light from both sides of the room.

Inside the apartment, the main items like the entire mechanical and infrastructural order, wall, ceiling and ground connections, material changes and door details are based on a continuous suture system. The rigidity of this system is balanced with the free and comfortable furniture and accessories.

The balance created by the contrast can also be seen in the choice of materials and colors. Natural stone covered walls, glass separators, painted glass and satin rust-resistant surfaces are in contrast with the old fashioned laminated parquet. The material balance is maintained by using familiar, warm, authentic and classic style accessories over clear, hard and serious looking material that dominates the overall appearance. A monochrome concept of color is used according to the client' s demand. The demand to create a life space which is casual, comfortable yet elegant is fulfilled by using as few colors as possible.

The lighting design, aside from being functional, is built to add the color of the environment with light. The necessary amount of light is provided with indirect lighting by also using spot lights pointing at paintings and objects. The light holders, chosen for their functionality, were specifically selected so that the light, not the holders, is the matter of attention.

Project Name: Snow White Villa
Design Company: helin & co architects

项目名称：白雪公主别墅
设计公司：天籁设计

纤细如丝，醇美如诗
POETIC AND ROMANTIC

本案于荷兰埃斯波海岸沿线，小山之巅，坐北朝南，周遭小岛林立。海岛风情，妙不可言，令人浮想联翩。

原基地浓荫避日，绿意盎然。本案空间，绿树当仁不让地占有一席之地。空间各开口、门户自然是依其地形，据其功能，投业主所好的综合考虑。

落日余辉下的门庭，倍显开阔。主出入口面朝夕阳，倍显妩媚。走出来，从庭院过小桥，到楼梯，直入洗浴室、厨房。足不出户，享尽生活之舒适，自然之清新。

混凝土结构，支撑的钢柱，立面的雪白瓷砖，轻盈的内饰，实木的橡木地板间以石灰石板点缀，共同为童话中的白雪公主打造着一个独一无二的晶亮世界。

The building locates on the top of a hill facing south, towards the archipelago spreading along the Espoo coast.

The old trees of the property played a big role in the designing process. The openings of the interiors were set by their location.

The main entrance locates through a closed outdoor atrium, facing the west towards the evening sun. The atrium can also be entered from the kitchen and via a bridge and a staircase from the sauna.

The concrete structure of the building was casted on the location over steel pillars and cladded with snow white plastered tiles. The surfaces of the interiors are light and the floors are mostly solid oak and limestone.

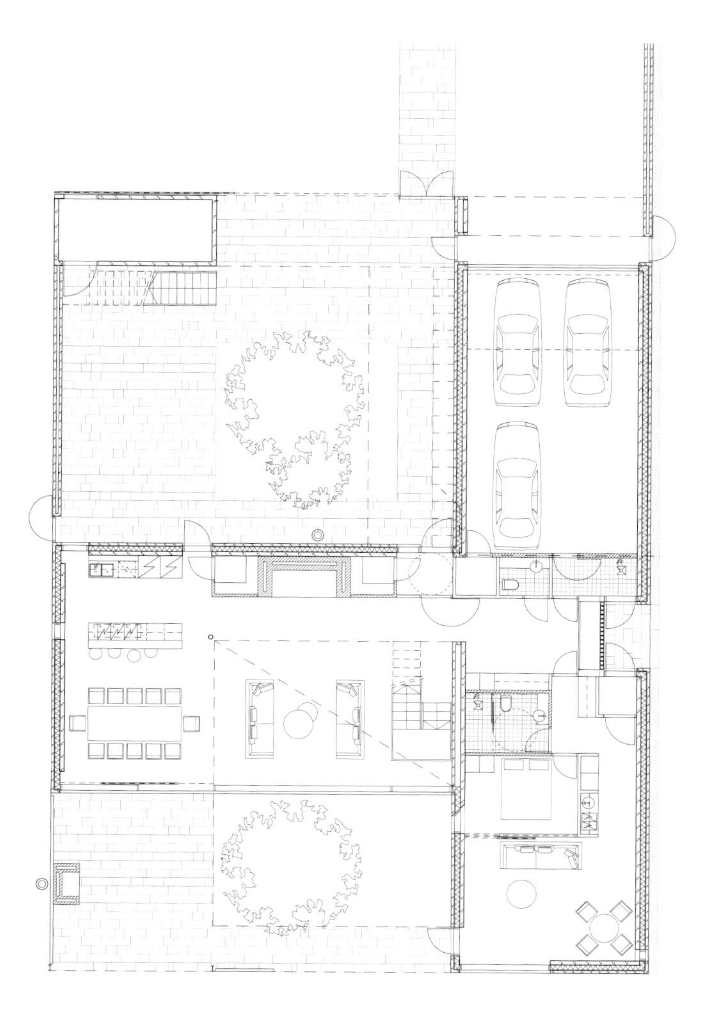

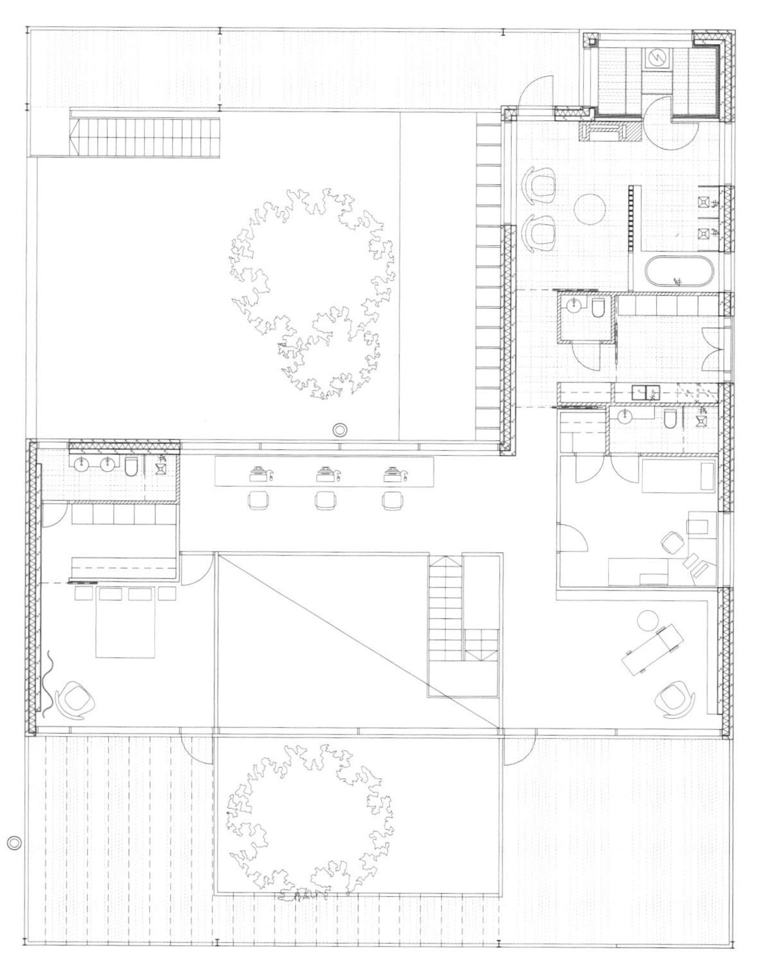

Project Name: Hollywood Hills Residence
Design Company: Bush Interiors
Photographer: Grey Crawford

项目名称：好莱坞山庄别墅
设计公司：布什室内设计
摄影师：格雷克劳福德

时间酿造出加州牧场的醇厚味道

CALIFORNIA PRAIRIE

应业主要求，重新设计后的建筑，清爽、现代、阳刚、简约、时尚。各主要公共空间的隔断得以移除，共同打造生活区域之恢宏气象。白色水磨石地板过厨房，经饭厅，走客厅，到达休息室及玄关。白色的纯洁，沙黄的温暖，天蓝色的安静创造着一个轻盈但依然富有变化的调色板。深色的胡桃木元素，是深度，是性格，也是对比。红色雕塑，颇有震撼，恰恰位于调色板留白处，聚焦着室内的视线。

餐饮组合台、茶几由本案设计及厂家共同打造，自成抽象美感。主生活空间物件铺陈，各有垂饰灯光照亮，以显其生命光华。盘门墙雕塑、各门户的舷窗设计、另有“克里姆特”灯具重复使用。主卧沙黄基调，另有白色、核桃木色精心点缀。墙体饰以手工作物、亚麻等面料。纸糊木框，以其旋转、滑动的姿态书写另类的曼妙，细打量，却是主浴的所在。大型的定制平台，静静地卧于一隅，那是朗朗的乾坤大气派。

The redesign of this mid-century home in the Hollywood Hills area was driven by the owner' s wish to create a clean, modern environment that would be minimal, masculine, and stylish. The goal was to open up the main public rooms to create one large open living area. The existing white terrazzo floors visually linked the kitchen, dining room, living room, lounge and entry. Pale tones of white, sand, and sky blue were introduced to create a light but varied base palette. Dark walnut elements were then added to provide depth, character, and contrast. This palette was intentionally reserved to provide a neutral foundation for the striking modern red sculpture that is the focal point of the interior.

The abstract forms of the dining set and coffee table were custom designed by Bush Interior and Silho Furniture. These pieces anchor the main living space. A pendant light fixture marks the position of each piece from above. The forms of the light fixtures by Le Klimt are repeated in a concrete-disc wall sculpture and more subtly in the porthole windows in the doors.

The palette of the master bedroom is a subtle study of white and walnut on sand. The bedroom walls are covered with grass cloth; the master bathroom is concealed by pivoting and sliding shoji screens. A large, custom platform bed is the focus and backdrop of the bedroom and gives the impression of a larger space.

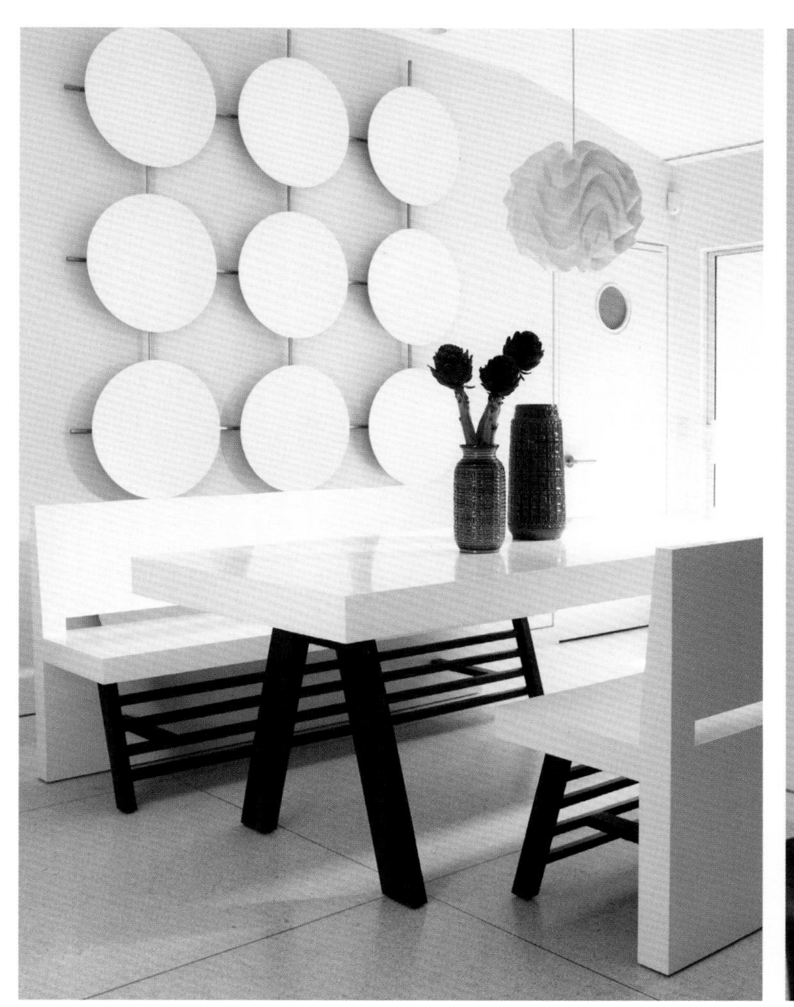

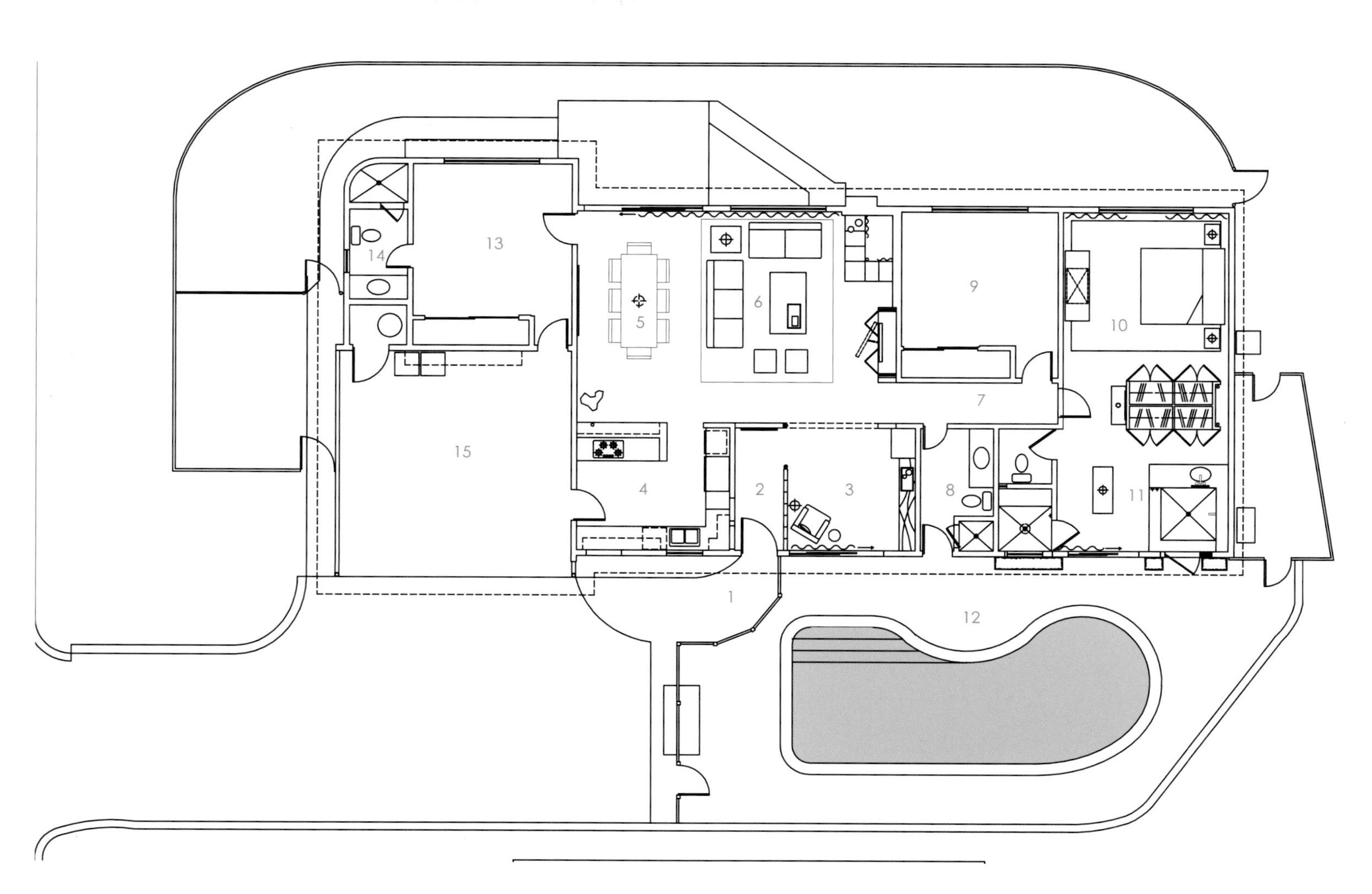
1
2
3
4
5
6
7
8
9
10
11
12
13
14
15

Project Name: Via Los Padres Remodel
Design Company: AB Design Studio, Inc.
Designer: Clay Aurell, Josh Blumer
Photographer: Dana Miller

项目名称：圣巴巴拉山麓老房改建
设计公司：AB 设计工作室
设计师：克莱 、乔什
摄影师：达纳·米勒

纯朴生活美学

LIFE CAN BE PURE AND PLAIN

圣巴巴拉山，风景秀丽，扬名天下。山下独有大片土地，生态自然。本案恰位于该地块边缘，远眺圣伊内斯山秀丽河山。其量体初建于 20 世纪 70 年代，典型的加州牧场风格。

几十年前，加州沿海房产蓬勃发展，圣巴巴拉山土地需求急剧增加。数十年前建筑依然不能把握时代脉搏，一如建于 30 年前的本案，平添些许沧桑，些许古旧。

但凡设计，总是承载着业主的殷殷期望。承其所待，重新设计后的本案，无论其外观还是气韵，均能彰显其经典内里及地段价值。瞧，当风雨走过几十载后，独立的凉台建筑，映照着气势恢宏的泳池，极富现代建筑特点。真乃新兴生活，家居所在。

Located amongst the scenic foothills of Santa Barbara, this residential property is located on the edge of a large undeveloped tract of land and enjoys a terrific view of the Santa Ynez Mountain Range beyond. The existing residence on the property was constructed in the late 1970's in a typical California Ranch style.

California's coastal real-estate values and a demand for developable land in Santa Barbara have dramatically increased over the past few decades. Many of the homes built on these prime properties years ago no longer retain or represent the current value of the land they now sit on. The current value of this particular property was no exception. The humble house built over 30 years ago was dated and it was beginning to show its age.

At the onset of the project, the owners of the property asked us to transform their existing residence into a home that would be worthy of this valuable and beautiful property. They wanted to upgrade the architecture of the house and improve the site and landscaping. The solution is a newly remodeled modern home with distinctive architectural features, a new free-standing pool cabana building and a grand swimming pool. The existing landscaping has been removed and will be replaced with a new design that will add to the new life of this property.

Project Name: Black White Residence
Design Company: David Jameson Architect
Designer: David Jameson

项目名称：黑白别墅
设计公司：大卫杰姆逊建筑师事务所
设计师：大卫杰姆逊

黑与白构筑晶亮住宅
THE RESIDENCE BLACK AND WHITE

本案砖石立面，如同洞穴，古风蔚然。而设计以主层为重，同时不经渲染，在其上新添另一楼层。

其空间映像如雅典卫城。光束铺面，四面现代化玻璃延伸于同一白色灰泥基座，相互映照并彼此界定。各光束自然融合，其内黑色核心，自然繁衍各自空间生命质感。细打量，各光束竟然为光打造之量体，真可谓千娇百媚，白天采光，夜晚闪光发亮，别样风情油然而生。

主层灰泥基座，上以玻璃立柱，并有开窗，俨然是室外山水林木的门户。临街的一面，长长的窗户映衬着笔直的树干，独坐于窗前，静观车水马龙，心中自是一片坦然。抬头望天，树如华盖，另有青天。踱步由里，透过玻璃幕墙，那是后园的风波律动。

Inhabiting the masonry shell of an existing house, this project engages the phenomenon of ruins and explores the idea of aperture. The design program called for renovating the main level and adding a second level with a significantly smaller footprint.

Alluding to the Acropolis, four modern glass temples emerge from a white stucco plinth as volumes of light, and define space between each other. These volumes of light are stitched together by a circulation core sheathed in black that extends to cradle each space. The volumes are instruments of light, gathering natural light to the interior during the day and glowing in the landscape at night.

The glass volumes of the upper level and windows incised into the main level stucco plinth are thought of as lenses to the landscape. Facing the street side of the site, long thin 'census' windows give focus to the measured cadence of tree trunks while editing views to passing vehicles. Above, the glass volumes provide panoramic views to the tree canopy and sky above. To the rear yard, one glass volume slices through the plinth to frame unencumbered two story views.

Project Name: The Cascading Creek House
Design Company: Bercy Chen Studio

项目名称：溪楼
设计公司：贝西震设计工作室

溪边的低碳大宅
THE MANSION LOW–CARBON

顾名思义，溪楼当然立于溪边。精巧设计，去其建筑量体的外观，赋其如德克萨斯州中部石灰岩“积水层”之映像。屋顶如盆地，极利雨水收集。另有光伏发电系统、太阳能热水板覆盖其上。所收集雨水自然流动，水源热泵，辐射循环或加热或冷却，协力打造微气候。还有地接地环路连接泳池、水景，成就空间立体热交换系统，最大限度地减少生活起居对电力、天然气的依赖，尽显环保手笔。

The Cascading Creek House by Bercy Chen Studio was conceived less as a house and more as an outgrowth of the limestone aquifers in the geography of Central Texas. The roof is configured so as to create a natural basin for the collection of rainwater, not unlike the vernal pools found in the outcroppings of the Texas Hill Country. These basins harness additional natural flows through the use of photovoltaic and solar hot water panels. The water, electricity, and heat which are harvested on the roof tie into an extensive climate conditioning system which utilizes water source heat pumps and radiant loops to supply both the heating and cooling for the residence. The climate system is connected to geothermal ground loops as well as pools and water features, thereby establishing a system of heat exchange which minimizes reliance on electricity or gas.

Project Name: House by the River
Design Company: Forma Design

项目名称：河府
设计公司：福马设计

通透大宅，八面来景
LANDSCAPE EXPOSURE MAXIMIZED

波托马克河河畔，前一层、后两层建筑量体，306 m² 超大空间，很遗憾地不能得河景之青睐。

楼上空间，各小房间依次排开，三卧室，二卫浴，厨房，餐厅，客厅。可谓极尽功能之划分，但主卧、厨房、餐厅却为河景所抛弃。楼下河景独好，但除了一小卫浴及客人房，空间利用如同地下室，可谓是个不小的浪费。

因此，空间改建时，尽可能缩小楼梯客间，左边生活、餐饮、娱乐空间得以扩大；阳台安装系列落地门。右边三小房间合成一大型主卧，其中睡眠、休息、化妆空间经纬分明，躺、卧、坐、行，抬眼静观河波涟漪。楼下，另辟家居空间，如书房、客房，并且扩大原有卫浴空间，以方便客人。改建后的空间，可谓八面来景，四面通透。

本案设计倍受推崇。2008 年 3 月《DC 现代豪华》杂志予以收录，同年获得国际室内设计住宅类金奖 (IIDA)。

This project was a total gut and renovation of a very modest 3 300 sf. mid-century tract house on a slope by the Potomac river. One story in the front, 2 stories in the back; a house that didn't really function well or even acknowledge the views to the river.

The upstairs was a series of small rooms jam-packed together – 3 bedrooms, 2 baths, a kitchen and dining room and living room, yet the master bedroom and the kitchen/dining area had no views to the river. The downstairs was not used – was one large storage room with a "basement" feel, yet with incredible views of the river, and a small guestroom and bath.

7217

DRESSED IN
BLACK
VALERIE MENDES

We opened the upstairs up as much as possible, creating one large living/dining/entertainment area on the left, and widening the opening to the balcony by inserting a series of French doors; combined all 3 bedrooms into one comfortable sized master bedroom suite, on the right, with a sleeping area, lounge area, dressing area and master bathroom that all had views to the river; and subdivided the lower level storage area into a family room and a study/guest bedroom, while enlarging the guest bathroom. Functionally the renovation opened the house to light and views from all sides, introduced color throughout (anything but beige, they said...) and now brings in the river views that become an integral part of the experience of being there.

This project was featured in the March 2008 issue of DC Modern Luxury magazine and won a Gold Award in the Residential Category from International Interior Design Association (IIDA).

Project Name: Eden View House
Design Company: MAP Architecture + Planning Limited
Indoor: 270 m^2
Outdoor: 227 m^2

项目名称：伊甸园大厦
设计公司：MAP 建筑策划
室内：270 m^2
室外：270 m^2

转折有致，一层一风景
DIFFERENT LEVELS, VARYING VIEWS

城市海景，尽拥空间。别致理念，楼梯独领风骚。通透玄关，沿楼梯宛转而上。上观楼上餐厅，其氛俨然；下看楼下起居，其为怡然。通透、开放、轻松，一切尽在本案空间。

Town house with sea views, the house has a half level change arranged around the stair case. The big design idea was to widen the entry hall, construct open riser stair to upper ground floor formal dining room opening the view to the lower ground floor living room. The house feels bigger, more open and easier to understand than all the other houses in the terrace.

VIKING

Project Name: Billson Kingussie House
Design Company: MAP architecture + planning limited
Size: 2 100 m^2

项目名称：比尔森别墅
设计公司：MAP 架构策划
面积：2 100 m^2

半山观景台，清风拂面来

ACROSS THE HILL DECK, COMES THE BREEZE

独立地段，超大面积花园洋房，独立车库，面面墙体开窗，海天美景，尽在内里空间。楼梯彰显空间雅致，夹层玻璃踏板，开放竖板星夜闪亮。开放式厨房，封闭式洗涤水槽。近 42 m^2 的主卧，附设开放式主浴，倍显空间恢宏。

Free standing Villa Lot House, 2 100 sft. in large garden; covered car port connects to the house; Windows on all sides of the house provide ample light and stunning sea views; the feature staircase uses laminated glass treads and open riser which glow at night. Open plan kitchen with closed scullery behind. Open plan master bathroom creates a master bedroom of 450 square feet in area.

Project Name: Ho(m)me
Design Company: K-Studio
Size: 180 m²
Materials: Sandstone, Steel Panel, American Walnut Flooring, Painted glass

项目名称："我"空间
设计公司：K 工作室
面积：180 m²
用材：砂岩、钢板、美国胡桃木地板、彩绘玻璃

镭射雕刻，制造迷离梦境

LASER CUTTING AND DREAMLAND

180 m² 空间，以石墙分为公共、隐密空间。其上嵌以壁炉，电视等功能设计。细打量，恍然间，墙体纵深，令人感叹，竟然有收纳厨房之功能。

厨房、餐厅、起居室一起纳入公共空间，其用材耐用、持久。凉石、钢材以激光切割，尽显哥特式风格与花边状表面镂空之可爱。中央位置，雕塑适时出现，原来是业主酷爱跑车、以空气动力学来体现。咖啡桌，如同端景，与另一端的DJ 台、酒吧遥相呼应。

脚步轻移，进入茧状空间。内设卧室、浴室、客厅，其布局开放。另有休息、盥洗、电视视听等空间。家具铺陈移动设计，应使用而作不同变化，或现或隐，尽显设计之精巧。五块玻璃墙板，以电子衣服挂杆串联，一旦连接，自然而形成更衣室空间。建筑与人，和谐共生。尽在"我"空间。

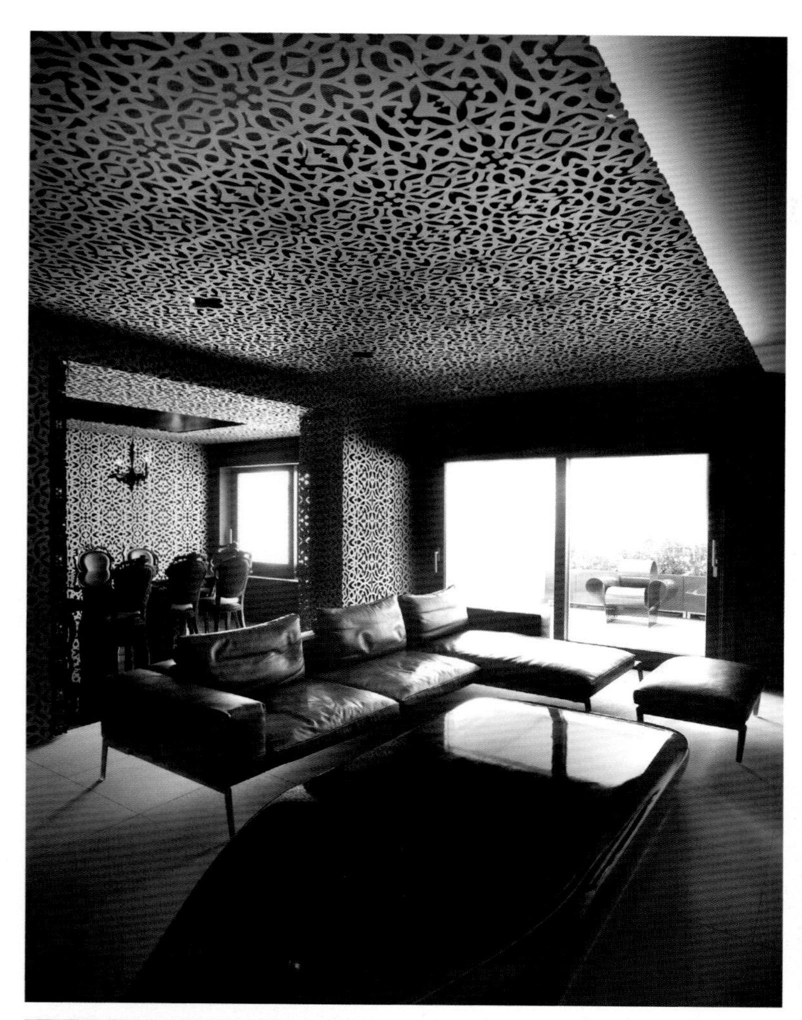

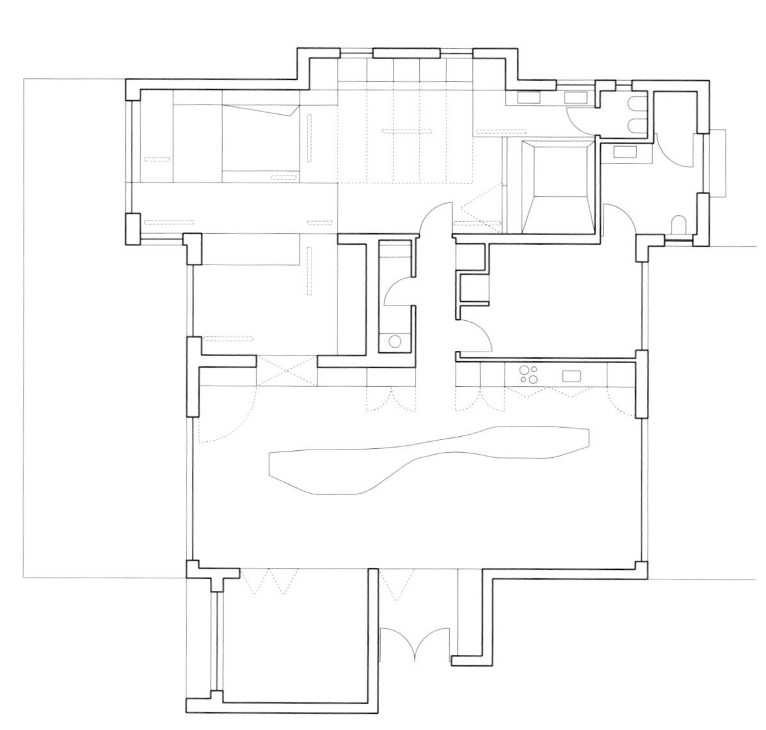

The 180-square-meter apartment is divided into 2 realms by a deep, stone-paneled wall containing the fireplace, TV and kitchen. The wall separates private space from public.

The kitchen, dining and sitting areas are incorporated into the public realm, characterised by durable materials such as cool stone and steel that is laser-cut in a gothic, lace-like pattern creating a perforated surface. His love of sports cars and fascination with aerodynamic forms inspired the sculptural element that progresses through the centre of the space, beginning at one end as a coffee-table and morphing into a fully equipped DJ booth and bar at the other.

Beyond the boundary a wooden topography cocoons the open-plan bedroom, bathroom and sitting area. This area needed to accommodate different programmes such as sleeping, washing, dressing, relaxing and watching TV. The response was to provide flexible, sliding components that re-shape the room according to the activity and can then be hidden. For instance a sliding, folding panel opens to connect the bath tub with the whole space. 5 glass wall panels slide into the space electronically pulling hanging rails of clothes and creating a dressing room.

Hom(m)e Tells a multi-narrative story of the symbiotic relationship between architecture and client, each enhancing the evolution of the other.

Project Name:870 UN Plaza
Design Company: Andre Kikoski Architect

项目名称：联合国大厦 870 号
设计公司：安德鲁建筑师事务所

质感取胜，成就品质豪宅
TEXTURE UTMOST

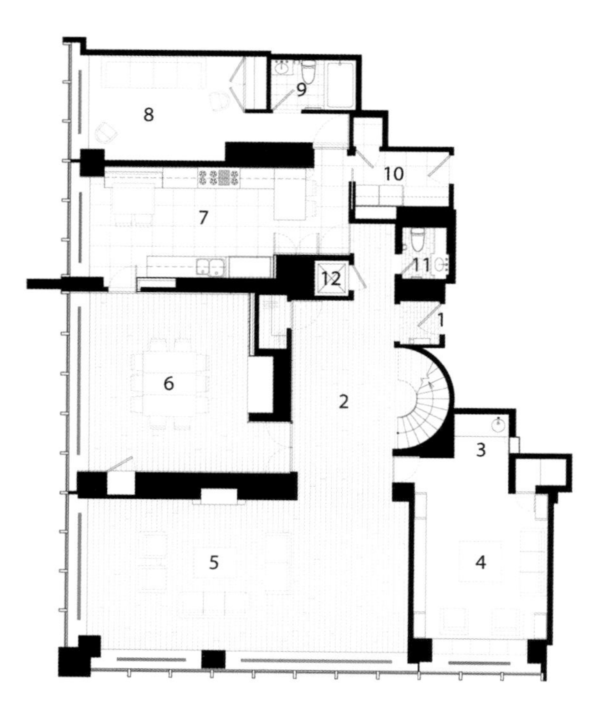

1
2
3
4
5
6
7
8
9
10
11
12

特殊的用材，考究的细部设计，是本案的气度，成就着本案418 m^{2}的与众不同，彰显着其在曼哈顿的现代主义色彩。各房间宽景设计，气势恢宏。贴心设计，尽在精彩娱乐，尽在其自然气韵，尽在其雅致日常生活。豪华的纹理用材共同铸就的调色板独特而低调。访客为之感到震撼，居家者为之舒心惬意。漆板墙体间以青铜雕像及花岗岩板塑造；手工制作的银色地板瓷砖，圆形的楼梯踏板呼应着墙上的皮革面板；到处运用的美国黑核桃木制品彰显着其优雅的品味。先进的材料，卓越的工艺，创造一个温柔、优雅的现代主义空间，那是业主的视野体现。

Exceptional materials and careful consideration of details distinguish this 4 500 sf. duplex in one of Manhattan' s great modernist tower. With its commanding views from every room, the apartment is a natural for grand and intimate entertaining. Our design highlights this attribute and emphasizes gracious daily living as well. The sumptuous textures and material palette are unique and understated. They distinguish the apartment for visitors, but also enhance the residents' everyday experiences. Walls of lacquer panels that are divided by statuary bronze inserts, Verde Maritaka granite, hand-made silver floor tiles, and circular stair set against a wall of leather-wrapped panels and extensive use of American Black Walnut millwork are just a few of the elegant custom finishes. The combination of these sophisticated materials and superior craftsmanship create a soft and elegant modernism in fulfillment of the owner' s vision for the space.

Project Name: Higgins Lake House
Design Company: Jeff Jordan Architect

项目名称：希吉斯湖滨苑
设计公司：杰夫·约旦建筑师事务所

双向叠加，坐拥双庭院大景观
THE ACHIEVEMENTS BY TWO PATIOS

密歇根州素有湖滨度假的传统。农舍小屋，田园风光。希金斯湖，全美最壮观的内陆湖泊，备受青睐。本案临湖而居，空间开敞、通透。其设计本着别样风光山色，大家共享之“博爱”理念，其烹调、饮食、体息区域划分明确，即便二十人入内，亦无丝毫拥挤、逼仄感觉，令人赞叹。

本案单栋独享，但地基狭窄。所幸，借设计之妙手，四个独立空间，相互共生，如同一高阔量体隐身于众房舍之中，无任何突兀之感。各空间采光、通风，静享山林、湖光美景。

主生活空间、主卧位于上层。次卧、客卧半隐设计。整个空间无任何量体之扩张，但气韵、内里尽显宏伟大气。主生活空间内的大型玻璃幕墙，极便于湖光山色的享受。沿南墙，更有玻璃设施。任其室外冬季酷寒难耐，室内自是一片暖阳。温暖的季节，独上露台，凭湖临风。

Many families in Michigan share a tradition of spending weekends and vacations in cottages adjacent to the state' s collection of lakes. Higgins Lake is considered one of the most spectacular of the inland lakes and the empty-nester' s who commissioned this cottage wanted to be able to share this resource with their large family and many friends. In order to do so, they requested a large gathering area for cooking, eating and relaxing and space to sleep twenty adults.

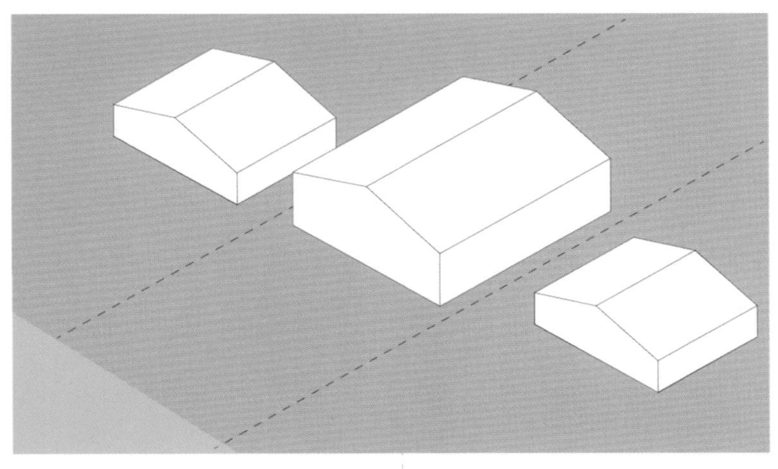

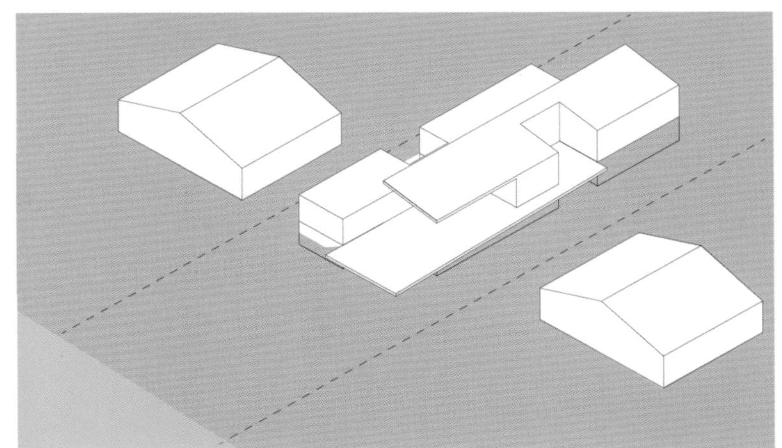

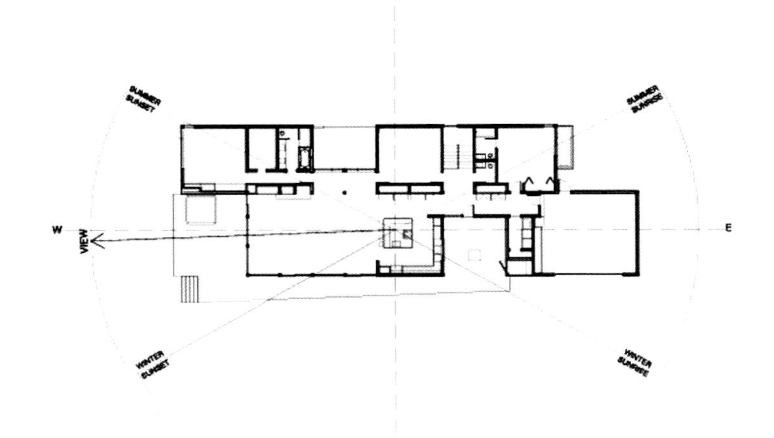

SUMMER
SUNSET
SUMMER
SUNRISE
W
VIEW
E
WINTER
SUNSET
WINTER
SUNRISE
S

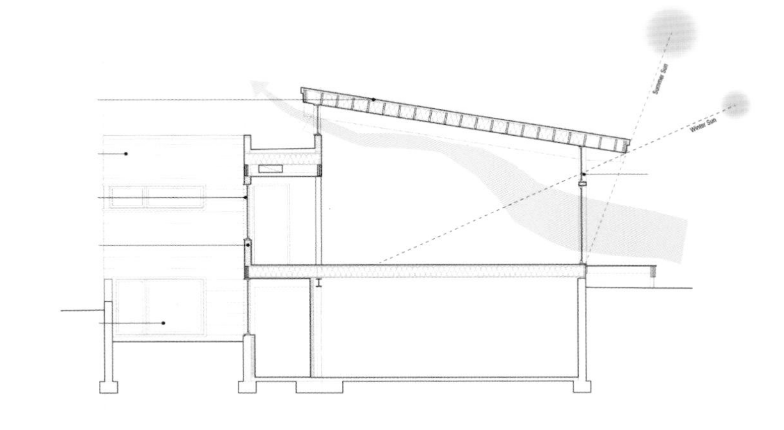

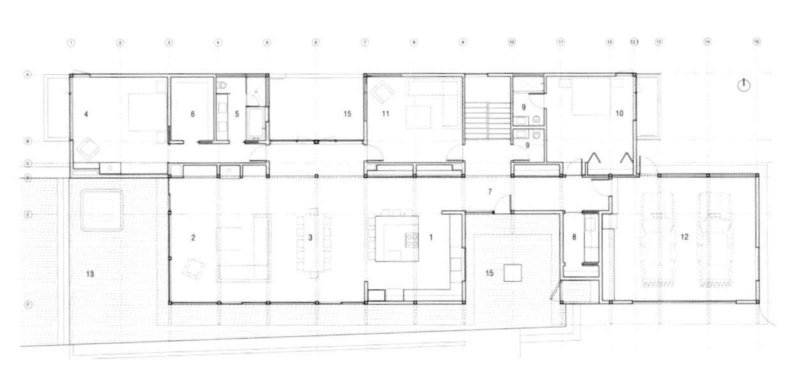

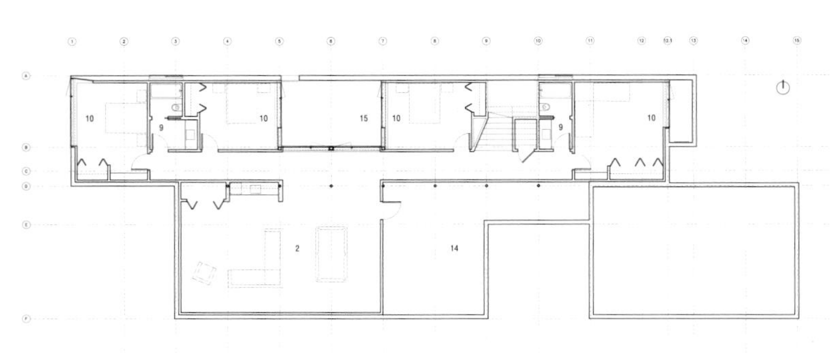

Given the relatively large size of the cottage compared to the neighboring houses, the central challenge of this project is fitting a large building onto a narrow site without overwhelming the neighbors, but still taking advantage of the views. Our solution locates the house in the center of the site and breaks the overall volume of the building down into four smaller volumes to better incorporate it into the surroundings. Additionally, this strategy brings substantial amounts of light and air into every room of the house while allowing for views out to the lake and surrounding forest. The living quarters and master suite are located above grade while the remaining bedrooms are partially buried to further diminish the apparent size of the house. The living quarters have a large expanse of glass facing the lake and additional glazing along the south side of the house to take advantage of passive heating in the winter. A large deck on the lake side of the house provides views while extending living space in the warmer months.

Project Name: HOUSE VK 1
Design Company: Gregwright Architect

项目名称：“海之蓝”
设计公司：格兰格怀特建筑师事务所

独特空间韵律，水泥质感大宅
TEXTURED CONCRETE

好的个案，流露业主品味生活的同时，也是设计师的名片。
本案主生活空间位于二楼，海洋宽景，中央庭院，各空间采光、透风尽然。
抛光混凝土地板，间以不锈钢材、自然灰石头、深色玻璃书写空间优雅现代格调，展日常独特生活。可调式照明，或朦胧，或舒雅，或大气，或小家碧玉，应空间使用要求，适时变化。

Conceived as the home for a young/trendy/elegant couple, this house is the perfect reflection of the client's brief.
The main areas of the house are raised and located on the first floor in order to take advantage of the beautiful views towards the ocean, leaving the complementary areas on the ground floor, spinning around a central courtyard that not only provides light and ventilation but also a beautiful view from every room on that level.
A very elegant and modern palette of polished concrete floors, textured concrete, stainless steel, with touches of natural grey stone and dark glass have conform the canvas for a unique lifestyle. The lighting design accents the spirit of the house on every room and determines a specific mood for each occasion by washing the different surfaces and creating a very rich fest of textures and shades.

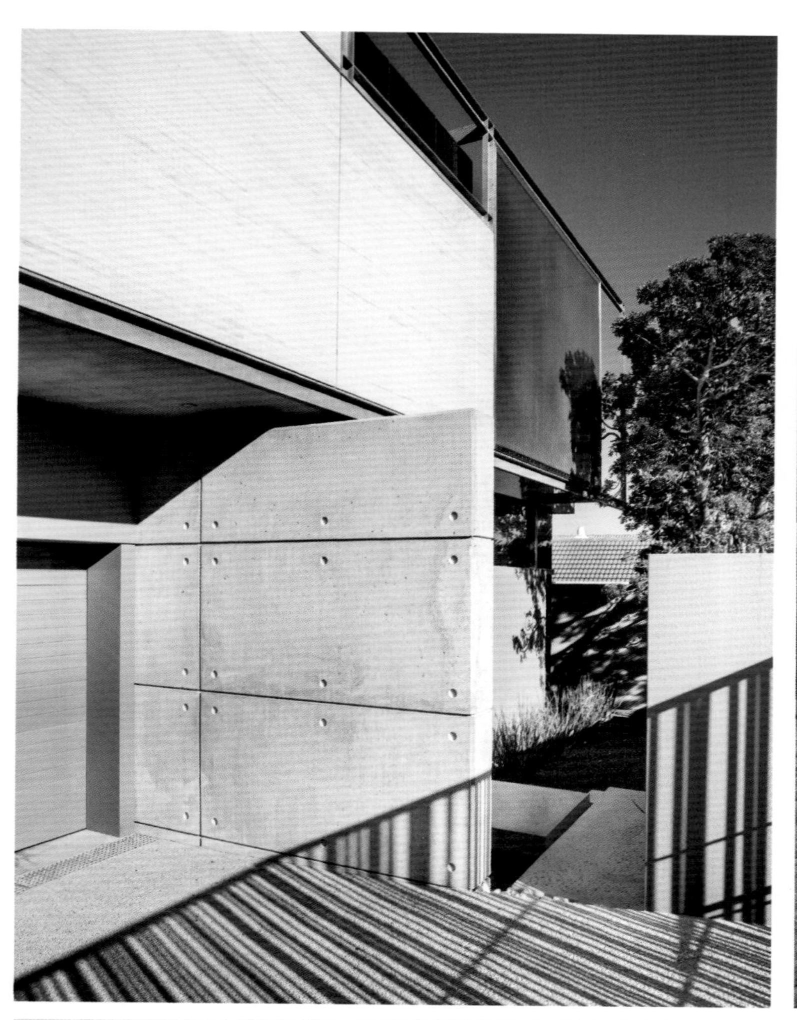

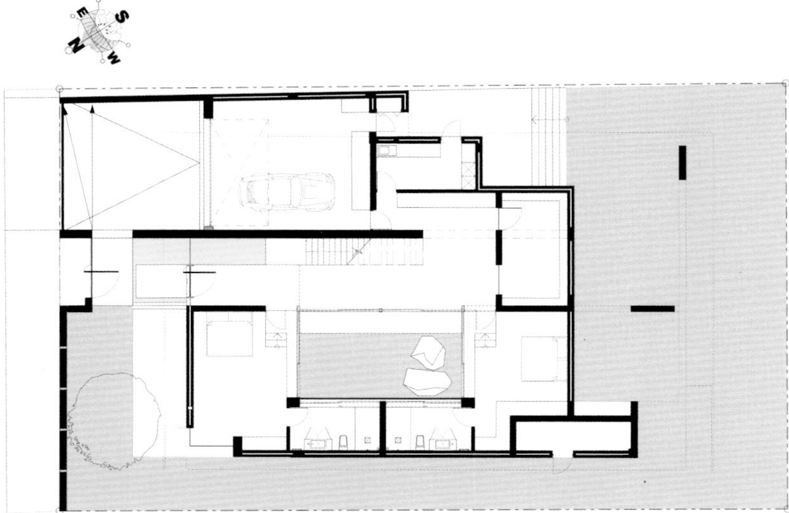
S
E
N
W

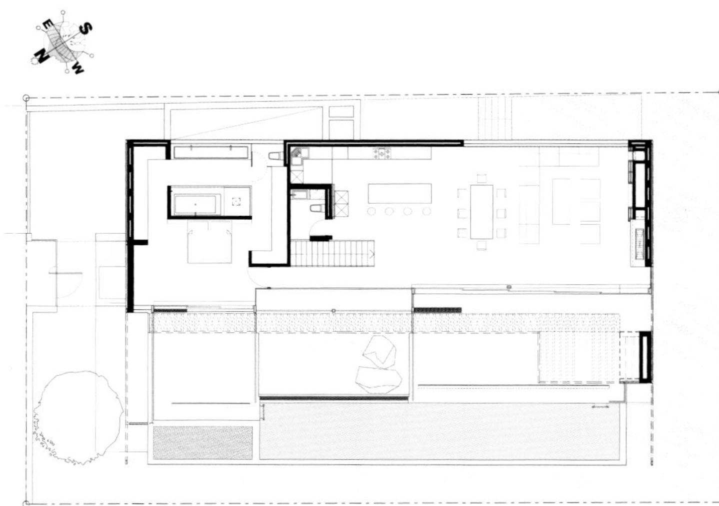
S
E
N
W

Project Name: Private House No. 2
Design Company: Lanciano Design

项目名称：悦己别墅 2 号
设计公司：莱斯安奴设计

美时美刻，与风景相拥
THE SCENERY RIGHT HERE

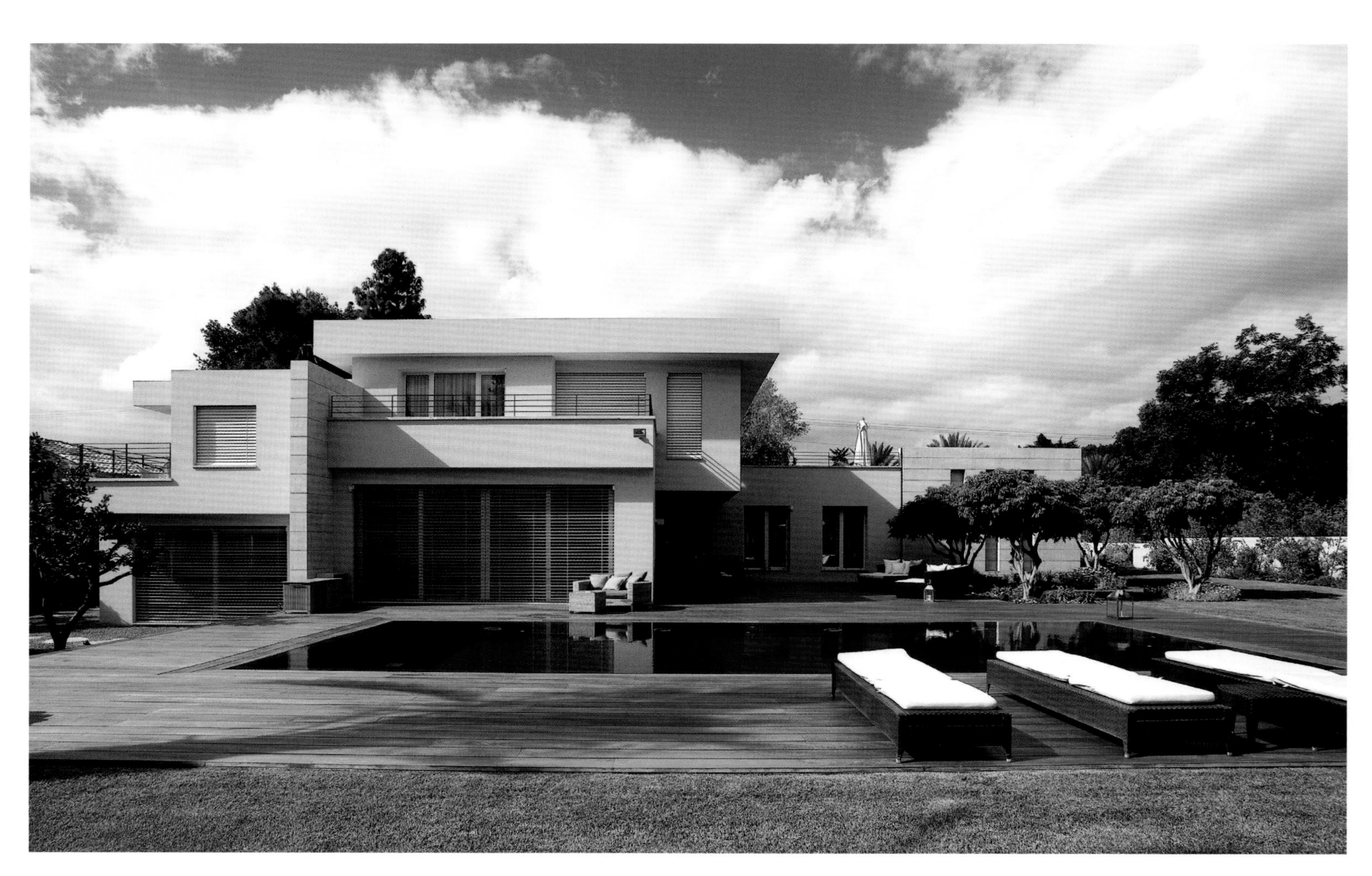

本案业主夫妇来自欧罗巴，品位高雅。妙手设计，打造高贵气度、时尚但依然内敛。

远观绿野苍茫，自然、建筑合二为一。一旦入内，双高空间，高阔量体，采光通风，彰显空间大气。

楼梯量体两两独立。扶栏玉砌，精致、典雅一如贵妇人腕臂上的手镯。

阅览室全高空间，位于主生活区域，书香四溢，尽显主人之品位。

主生活空间，无墙体之阻隔，视线游动，真是花园中有家居，家居在花园里。

卧室延续开阔气势。木作装饰墙正对着床区。木香温润，安然入眠。细打量，墙后另有天地——步入式柜橱、卫浴。

卫浴自成一方世界。性质、功能截然不同的两种铺陈，面料地毯、水槽竟然你中有我，我中有你，并且合二为一攀升至天花顶端。

光照的夜晚，繁星闪闪，神秘夜，浪漫风。

This house was built for a European couple with a very refined taste; we were intrigued with the opportunity to design a house that reflects beauty & sheer elegancy and yet "shy" in a way.

The house from the outside seems to grow gradually from the landscape blending quietly with the beautiful garden that surrounds it. Once you come to the entrance

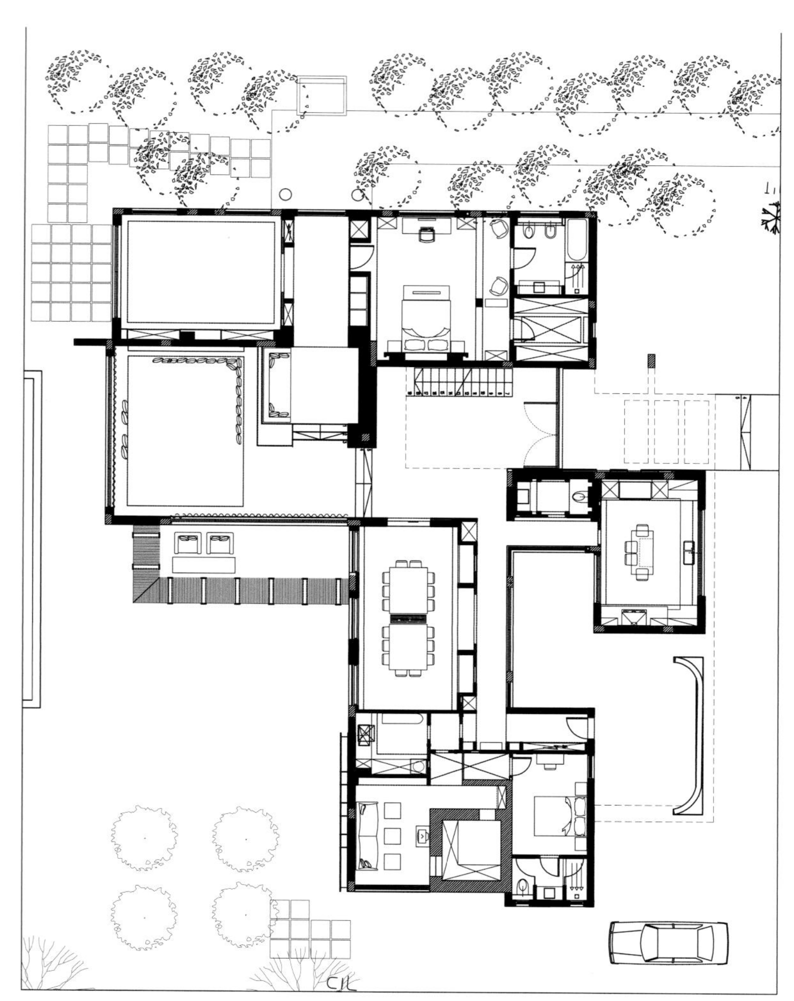

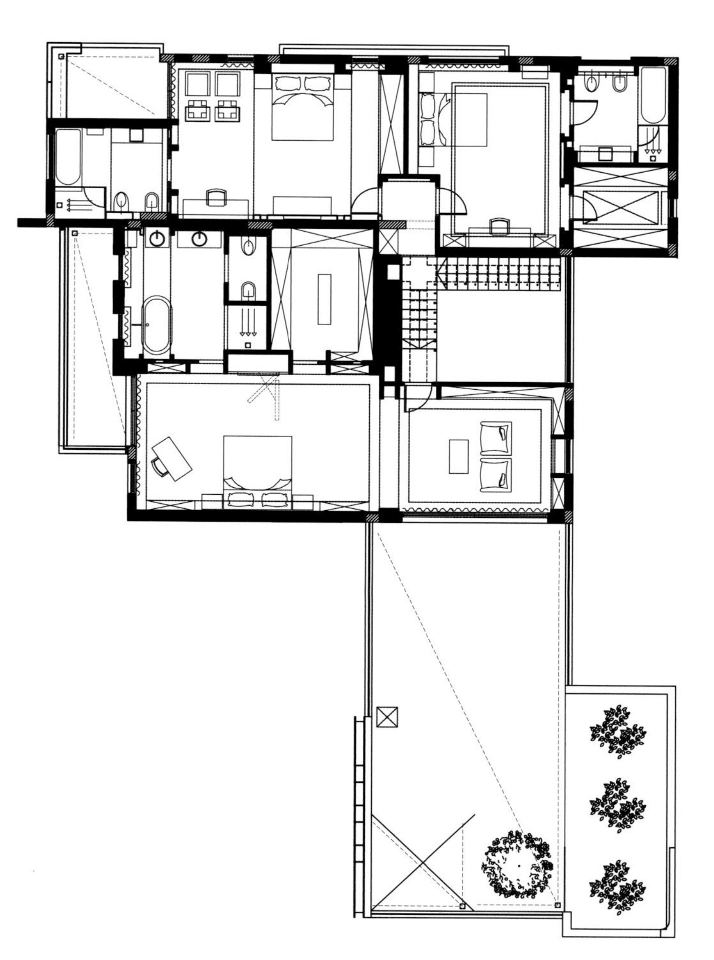

hall way the double space ceiling with the big window above allow the light to give its impact on the walls filled with art.

The stairs case is disconnected from the wall and looks floating in the space, the banister was designed as a jewelry looking like an expensive bracelet.

We created a big library at the main launch, using its full height to create maximum strength, giving the height its utmost importance.

The launch was completely opened with wall to wall windows to allow the garden & the house to unite.

The bedroom is fairly large; we placed opposite the bed a carpentry wall that hides behind it the walking closet & the bathroom.

The bathroom was designed as one space but we wanted to give the beautiful stand alone bath its own volume. There for we specified that area with a carpet designed on the floor going up to a designed sink that follows the carpet and climbs up to the ceiling.

The lightning at the house is really outstanding and at night it gives the house a mystic & romantic atmosphere.

SIMPLE AND PURE 简素

Project: Godoy House
Location: ZAPOPAN, JALISCO, MEXICO.
Surface: 302 m^2
DESIGN FIRM: HERNANDEZ SILVA ARQUITECTOS
DESIGN TEAM: ARQ. JORGE LUIS HERNANDEZ SILVA
ARQ. FRANCISCO GUTIERREZ P.
ARQ. DIANA QUIROZ CHAVEZ
ARQ. BELEN ALDAPA OROZCO.
PHOTOGRAPHY: CARLOS DIAZ CORONA

项目名称：Godoy House
项目地点：墨西哥
项目面积：302 m^2
设计公司：HERNANDEZ SILVA ARQUITECTOS
设计团队：ARQ. JORGE LUIS HERNANDEZ SILVA
ARQ. FRANCISCO GUTIERREZ P.
ARQ. DIANA QUIROZ CHAVEZ
ARQ. BELEN ALDAPA OROZCO.
摄 影 师：CARLOS DIAZ CORONA

悬浮宝盒
FLOATING TREASURE BOX

这间房子位于市外的一个私人住宅区。房子所处地势平坦，东面和南面都是其他住户，西面一排长长的树木把房子与街道分隔开来。北面是一个角位，房子的两个入口都在这里：一个是供人使用，令一个供车使用。房子由两个看似浮动的主体构成。左面的主体犹如一个被灰色石头覆盖的大盒子，漂浮于街道之上，车子便可以停泊在大盒子下方的空间里。右面的主体透明且轻盈，比左面的更大更高，把人行入口暴露出来。这个主体的北面和西面是两排大窗户，分别用白色的铁网覆盖，从而阻挡炽热的阳光。毕竟这间房子所处的方位是很难在城市内找到的。

入口旁边的树木使透明主体有延伸的感觉。其前方覆盖着铁网，阻隔西面的热气。后方是一面垂直的支撑墙。两面高墙水平倾斜，形成了高达 11 m 的巨大天花。天花横跨花园与泳池上方，并支撑着位于主体二层的主人房，从而使主人房具有开阔的视野，俯瞰被树林环绕的花园。

因为房子有一个半地下室的关系，入口平台被升高了。房子的内部完全开放，从庭院的入口可以清晰看到，房内的墙体几乎面向同一方向。而庭院的入口是空气流通的交汇处，一座垂直的桥和一座透明的桥连接了房子一楼的两头。

房子一楼长长的折叠式窗户把室外庭院融入室内，使得房子的社交空间成为与庭院和泳池融合的一个露台。这样的设计可以让人很好地享受瓜达哈拉哈的清凉天气。

二楼的饭厅俯瞰一楼露台，厨房和工作室则与庭院相通。地下室建有两座楼梯，一

22

22

座作日常使用，另一座则把房子的社交空间和私人空间串联起来。
屋顶的前方是平的，后方则遵从建筑管理稍稍倾斜，并覆盖有黄色的上胶陶瓷。房子的前面基本平滑，地板铺装采用大理石和原木。由于房子的木结构颜色都偏深，因此设计师应用许多白色的钢材元素，使房子更明亮。

The house is located in a private neighborhood outside the city, the land is flat and located in a corner, with neighbors to the south and east side, the west contains a large line of trees that separates the house from the street and takes a turn into the north where both entrances lie: one pedestrian and another one for cars, there are two bodies floating, the one on the left is a large box covered in gray stone that levitates over the street opening a gap where the cars are stored, the second body is transparent and light, larger and higher, which shows the pedestrian entrance, it is a volume of windows to the north and west sheltered by a white steel lattice, obeying the relationship with the sun, as this orientation is extremely hard in the city.
The transparent volume at the entrances expands by the side of the trees, the lattice covers its front face and is secured to a vertical wall at the back, which eventually blocks the heat from the west, in the back, two walls rise and bend horizontally to create a great flying ceiling of 36 ft. (11 mts.) over the garden and pool, supporting a second floor where the master bedroom is located so that it gets a complete open view to the garden which is surrounded by woodland.
The entrance platform is elevated because the house has a semi-basement, thereby it allows to generate different uneven levels opening the flow in several directions, the inside is completely open, with almost all the walls in the same direction, this is quite evident from the courtyard entrance, which is a well it double height gap, where the main circulations converge, one vertical and the other a translucent glass bridge connecting the two ends on the first floor.

22

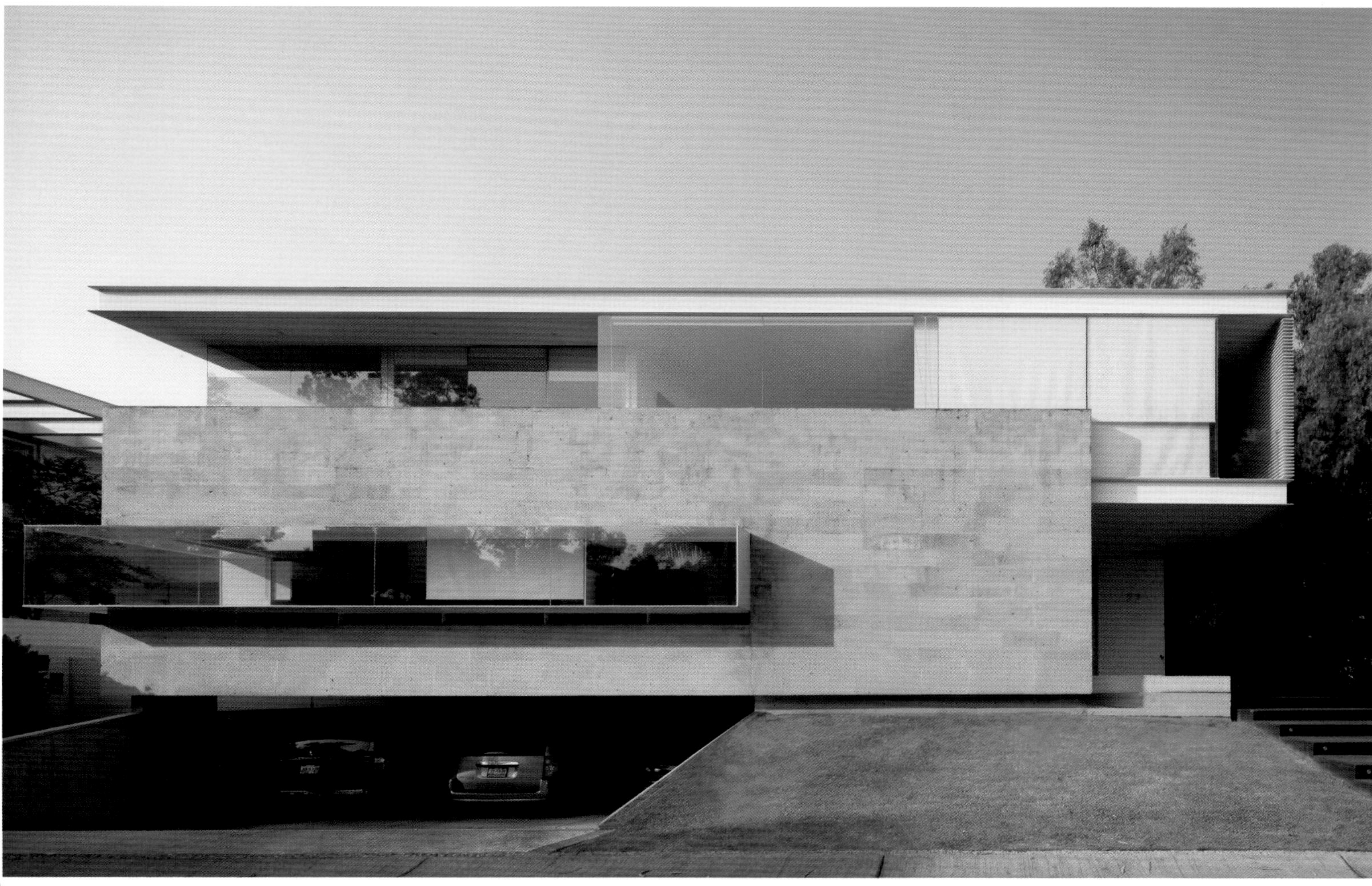

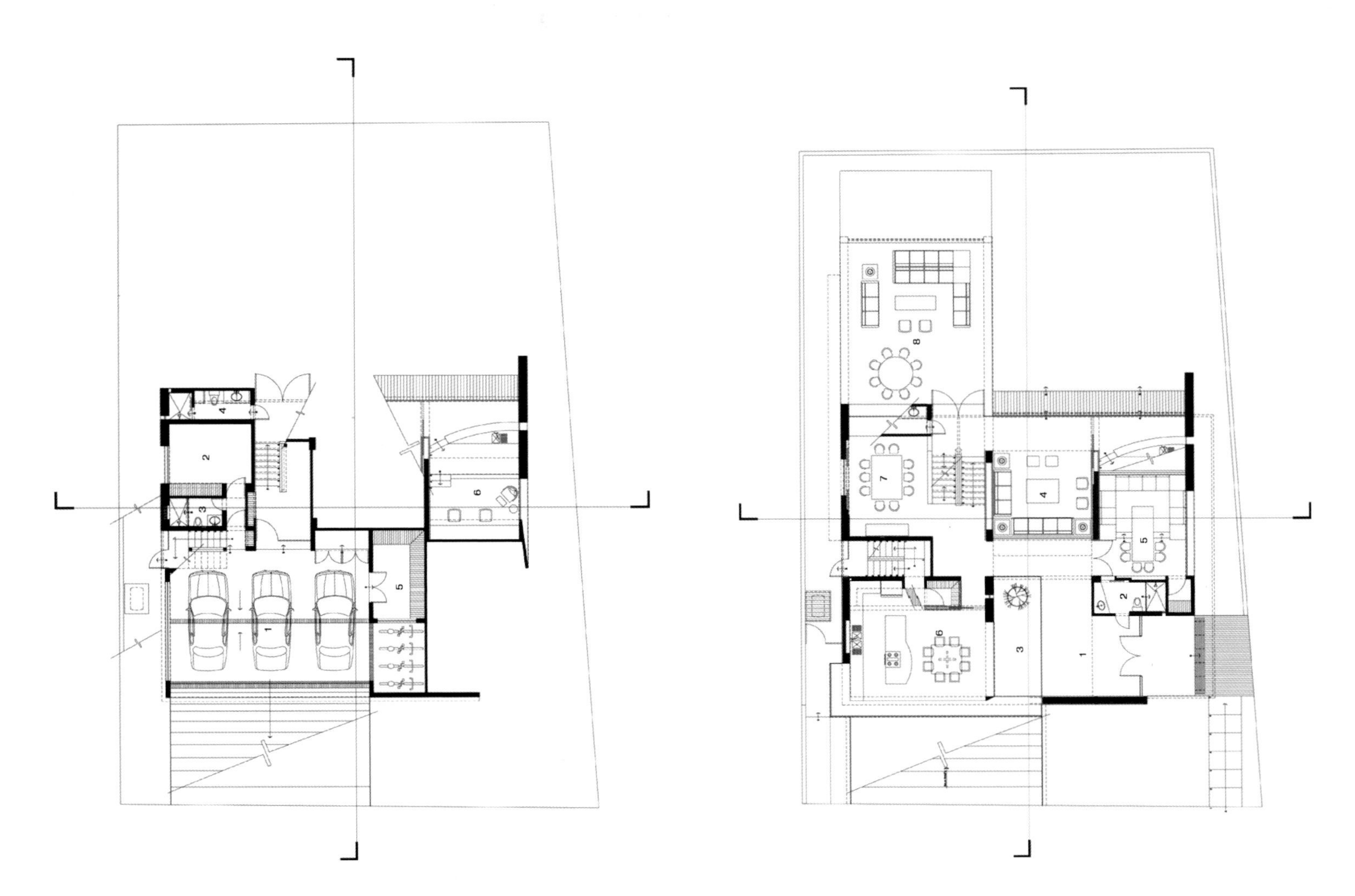

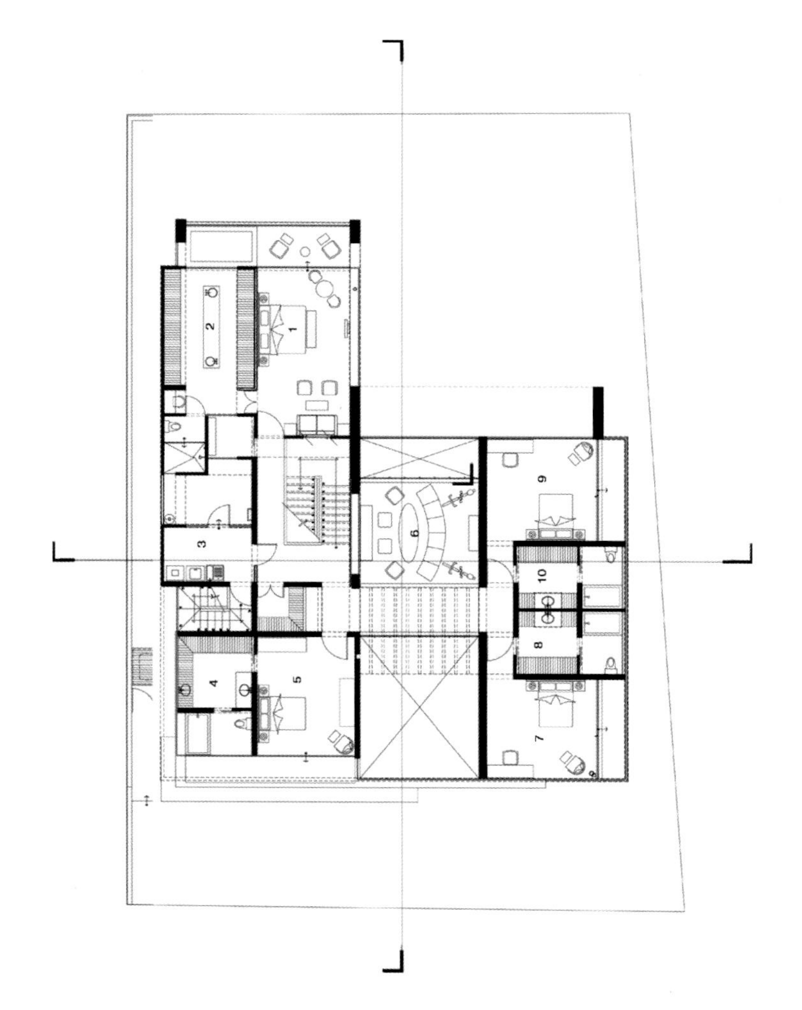

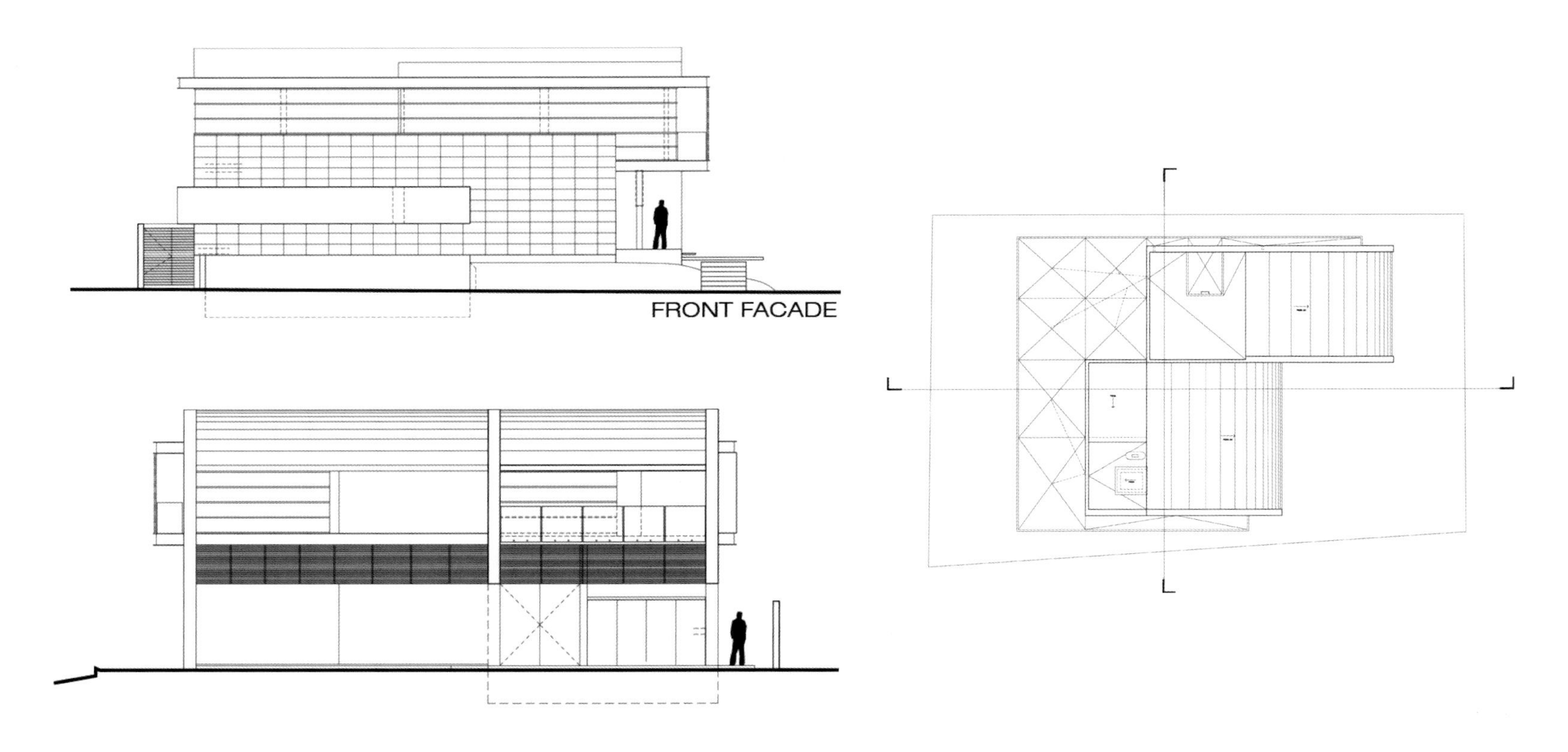
FRONT FACADE

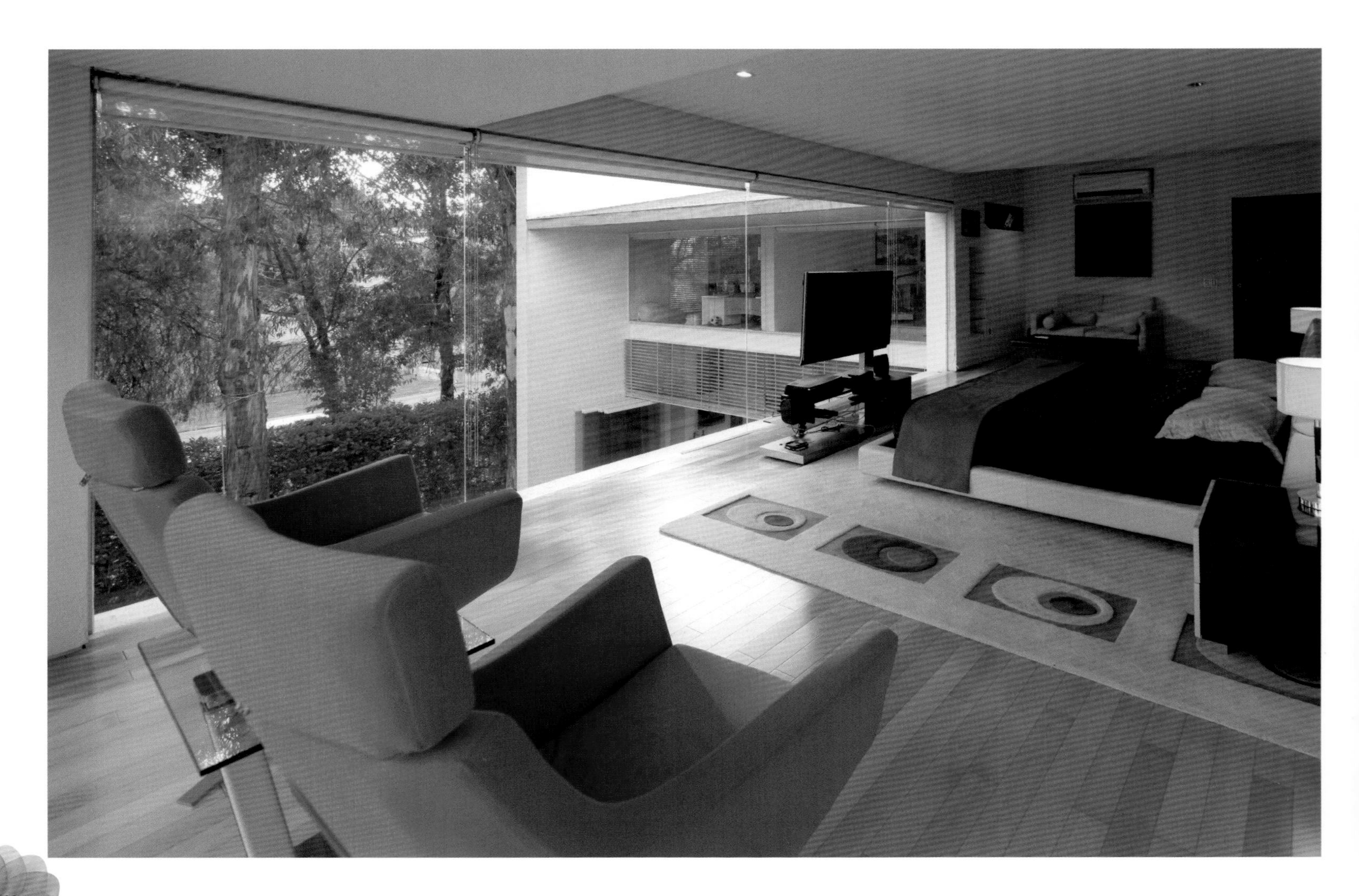

The house structure is steel, it was built during a situation where this material was at a low price, allowing to build an almost floating house.

The house opens with a long and folding window system in the background, integrating the garden into the interior, making social spaces in a fully integrated large terrace with a garden and pool, this allows to live the cool and privileged Guadalajara' s weather, the dining room looks out above the terrace and communicates above the garden with the kitchen and studio, the two stairs set off from the basement to the second floor, one leads to the services and the other weaves the social and private spaces of the house.

The roofs are flat on the front but slightly declined on the back due to construction regulations covered with coated yellow ceramic, the walls are mostly smooth, floors are marble and wood, the carpentry is all in dark colors, using white steel on many elements of the house.

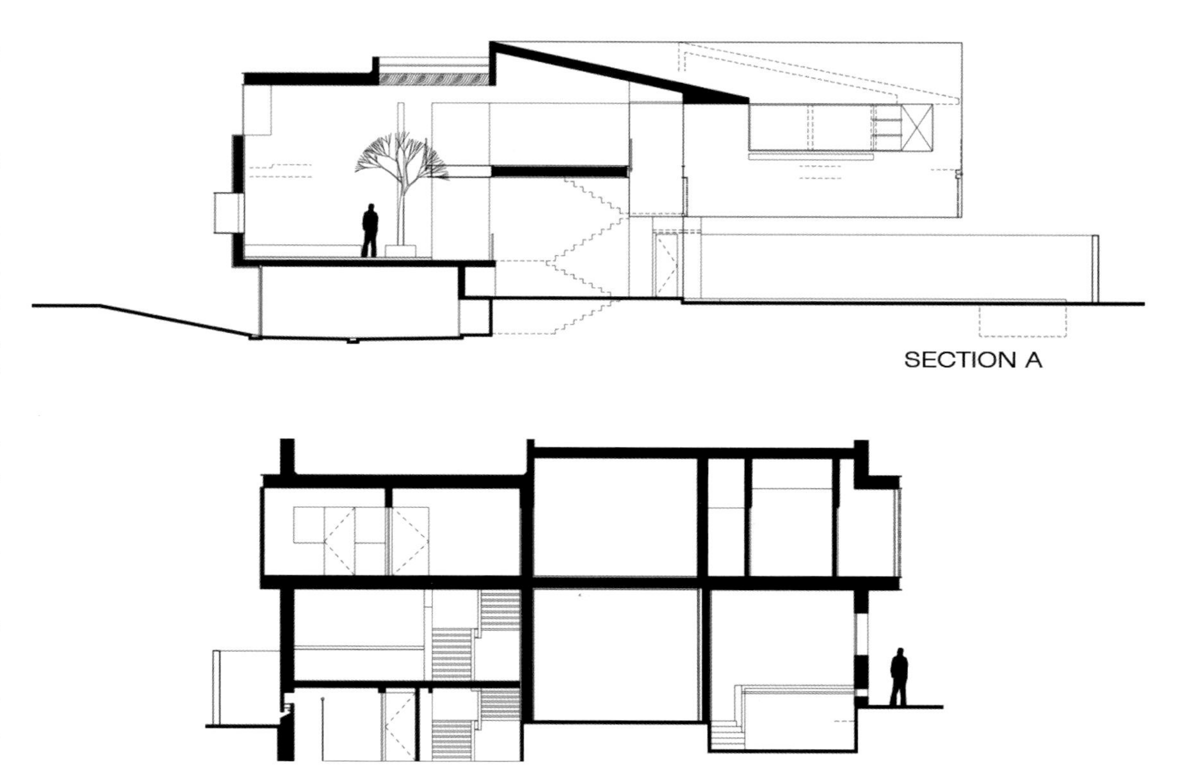

图书在版编目（CIP）数据

简美 . 2 / 黄滢主编 . – 武汉 : 华中科技大学出版社 , 2012.3

ISBN 978-7-5609-7772-0

Ⅰ . ①简… Ⅱ . ①黄… Ⅲ . ①住宅 – 室内装饰设计 – 图集 Ⅳ . ① TU241-64

中国版本图书馆 CIP 数据核字 (2012) 第 040718 号

简美II 黄滢 主编

出版发行：华中科技大学出版社（中国 · 武汉）

地　　址：武汉市武昌珞喻路1037号（邮编430074）

出 版 人：阮海洪

责任编辑：黎若君　　责任监印：秦英

责任校对：段园园　　装帧设计：百彤文化

印　　刷：利丰雅高印刷（深圳）有限公司

开　　本：965 mm × 1270 mm　1/16

印　　张：20.5

字　　数：164千字

版　　次：2012年5月第1版 第1次印刷

定　　价：298.00元（USD 59.99）

投稿热线：（020）66638820　　1275336759@qq.com

本书若有印装质量问题，请向出版社营销中心调换

全国免费服务热线：400-6679-118 竭诚为您服务